Berger · Aroma Biotechnology

Springer
Berlin
Heidelberg
New York
Barcelona
Budapest
Hong Kong
London
Milan
Paris
Tokyo

Ralf G. Berger

Aroma
Biotechnology

With 61 Figures and 48 Tables

 Springer

Professor Dr. Dr. Ralf G. Berger
Universität Hannover
Institut für Lebensmittelchemie
Wunstorfer Straße 14
D-30453 Hannover, Germany

ISBN 3-540-58606-7 Springer-Verlag Berlin Heidelberg New York

Cip data applied for

© Springer-Verlag Berlin Heidelberg 1995
Printed in Germany

Typesetting: Data-conversion by M. Schillinger-Dietrich, Berlin
SPIN:10126646   52/3020-5 4 3 2 1 0 - Printed on acid-free paper

Preface

Food scientists, and even food microbiologists, frequently associate the mention of microorganisms with substandard quality or spoilage. However, empirical food biotechnologies are the roots of modern food technology, and products, such as beer, wine, vinegar, bakery goods, and fermented vegetables, milk, meat and fish owe their typical aroma profiles to the metabolic capability of numerous microorganisms. Recent advances in fungal and plant biotechnology, enzyme technology, genetic engineering, bioprocess monitoring, and product recovery techniques pose novel potential opportunities for the biotechnological generation of aromas. Those interested in the field must search through several conference proceedings and many journals of enzymology, plant science, mycology, chemical engineering, sensory science, molecular biology, and more to obtain something like an overview.

The present introductory text is intended to provide guidance through this rapidly proliferating area of food biotechnology. The first chapter deals with classification, functions, and bioactivities of aroma compounds. The second chapter is devoted to aroma aspects of the traditional food biotechnologies that are reconsidered as starting points of future developments. After a chapter on the multiple motives for aroma biotechnology, some practical information is covered for beginners, followed by separate discussions of microbial de novo syntheses, biotransformations using whole cells, and using enzymes. A chapter on genetically altered catalysts has been added to provide a picture of possible aroma applications, independent of future food regulation. Plant catalysts and a detailed outline of process design, monitoring, and aroma separation are treated in further chapters. In the closing chapters, special emphasis is put on process improvement and practical applications.

Every effort has been made to stress the most recent developments and to cite the most up-to-date literature. Key references are included for each subject area to facilitate access to information published before 1990. The attentive reader will note that aroma biotechnology, though based on the same fundamentals, cannot rely on the same broad experience as do other disciplines of biotechnology. Recipes how to obtain any given aroma compound in space/time-yields needed for a commercial process are not available. It can be expected, however, that aroma chemicals will become more and more accepted as another class of fine chemicals amenable to the biotechnological approach. The synthetic potential of the biocatalyst will ultimately determine, if a bioprocess or chemosynthesis is to be preferred.

Hannover   Ralf G. Berger

Contents

1 Aroma Compounds in Food

1.1 Aroma, Flavor, and Fragrance Compounds

The consumption of foods and beverages is inseparably linked to the stimulation of the human chemical senses, odor and taste. The sensation of odor (smell) is triggered by highly complex mixtures of small, rather hydrophobic molecules from many chemical classes that occur in trace concentrations and are detected by receptor cells of the olfactory epithelium inside the nasal cavity. The nonvolatile chemical messengers of the sense of taste interact with receptors located on the tongue and impart, though not limited by polarity or molecular size, four basic impressions only: sweet, sour, salty, and bitter. Also perceived inside the oral cavity, but transmitted to the brain by nonspecific and trigeminal neurons, are pungent, cooling and hot principles.

In scientific Anglo-Saxon usage all sensory (odor, taste, color and texture) attributes of food have been classed under the general term 'flavo(u)r'. Webster's Dictionary, however, defines flavor as 'the blend of taste and smell sensations evoked by a substance in the mouth', which would exclude color, texture and all impressions noticed before and after the phase of intake. This apparent confusion of physiologically separated entities and food related functionalities has resulted in an often inconsistent use of terms (Parliment and Croteau, 1986; Schreier, 1988; Bauer et al., 1990). For the sake of readability the following terms will be used as synonyms in this text: volatile flavor, odor(ous) compound, and aroma(s). Fragrance substances, used in perfumes, cosmetics, and toiletries, are distinguished from volatile flavors mainly by the different range of applications.

The first comprehensive list of volatile compounds in food comprised a few hundred constituents (Weurman, 1963). With the advent of modern instrumental analysis, particularly coupled gas chromatography-mass spectrometry, the number of identified compounds exploded, and the current compilation (Maarse and Visscher, 1992) contains more than 6200 entries from about 400 different food sources. The sensitivity of complementary, classical spectroscopy (high resolution nuclear magnetic resonance, Fourier transform-infrared spectroscopy) has been improved during the last few years, and has thus contributed to the structural characterization of aroma compounds. In spite of these achievements, the detection of new volatile flavors is getting more and more difficult. Extrapolating from the number of facultative aroma precursors, from confidential industrial results, and from recently published progress it has been estimated that up to 10 000 volatiles may be present in food. The greatest complexity of the flavor profile is found in products that, after a fermentative liberation of aroma precursors, have received additional thermal or other technological treatments: more than 500 volatiles have been reported for coffee, wine, beer, rum, black tea, and cocoa. Not surprising, the highly aromatic variants of these materials are commonly classified as luxury food. The figures also indicate a considerable overlap of many volatiles in foods with independent flavor profiles. This raises the question of the quantitative distribution of the single components in the mixture. Many separation and enrichment procedures have been developed culminating in the use of stable isotope labeled internal standards for distillation/extraction (Preininger et al., 1994), but each of the standard techniques shows different recoveries across the broad range of polarities and boiling points. Headspace concentrations are relevant for many applications, but do not permit an accurate calculation of absolute contents in nonfluid food matrices.

1.2 Character Impact Components

The interest of many researchers has now shifted to the determination of the actual contribution of single constituents to the over-all flavor of a product. The sensory importance, as a rough rule and expressed by odor units ('aroma value'), depends on the active concentration and the reciprocal of the respective odor recognition threshold, as determined experimentally, preferably in the food matrix. The weak point of this concept is the unpredictable extent of interaction of the flavor molecules with each other and with the other food constituents. A partial solution to the problem is coupled gas chromatography-olfactometry (GC-O), also called 'sniffing-gas chromatography'. Serial dilutions of a flavor extract are submitted to gas chromatographic separation, and the separated components are then detected physico-chemically and, in a parallel effluent of the separation column, by human panelists. This method, adapted from traditional sensory analysis, can be applied for the determination of odor thresholds in air as well as for the assessment of individual contributions of separated compounds (Berger et al., 1983; Acree et al., 1984; Schieberle and Grosch, 1987). Some psychophysical and other drawbacks of gas chromatography-olfactometry are obvious, but laboratory practice and first patent applications have clearly demonstrated the potential of the method. Recently accumulated experience with GC-O has confirmed flavorist's knowledge: Major constituents, such as carboxylic acid esters (fruity), benzaldehyde (stone fruit flavors), or vanillin may determine an over-all flavor; however, more frequently trace and ultra-trace compounds with low odor thresholds and actual concentrations often below the instrumental detection limits play a significant role. These volatile key compounds, usually addressed as 'character impact components', have received particular attention (Emberger, 1985). The common structure elements of the methoxyphenols or other, heterocyclic examples of Fig. 1.1 sug-

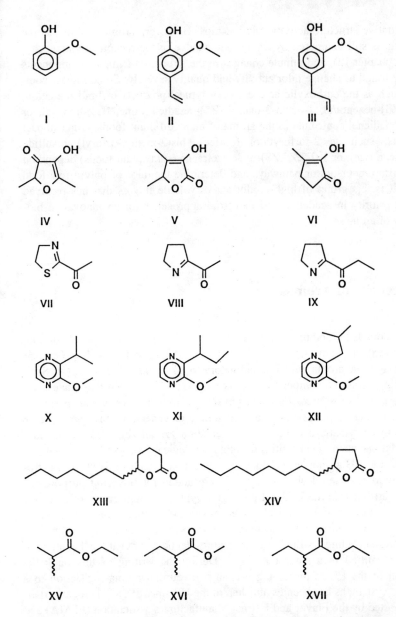

Fig. 1.1. Low odor threshold constituents of food I = Guaiacol, II = 4-Vinyl guaiacol, III = Eugenol, IV = Furaneol, V = Sotolone, VI = Cyclotene, VII = 2-Acetyl-2-thiazoline, VIII = 2-Acetylpyrroline, IX = 2-Propanoylpyrroline, X = 2-Isopropyl-3-methoxypyrazine, XI = 2-sec-Butyl-3-methoxypyrazine, XII = 2-Isobutyl-3-methoxypyrazine, XIII = 5-Dodecanolides, XIV = 4-Dodecanolides, XV = Ethyl 2-methylpropanoates, XVI = Methyl 2-methylbutanoates, XVII = Ethyl 2-methylbutanoates

gest quantitative structure-activity relationships. However, 'aroma design', unlike drug design, is still hampered by limited knowledge of the receptor(s) stereochemistry (cf. chapter 12), and minute changes in the structure of impact components are usually found to change odor activity and quality severely. Some impact components, such as the carboxylic acid esters or typical products of lipid oxidation, such as (Z3)-hexenol/al, 1-octen-3-one, 1-(Z5)-octadien-3-one, (E2)-nonenal, or (E2,Z6)-nonadienal contribute to the aroma of many different foods. Other impact components, such as 2-furfurythiol (coffee), bis(2-methyl-3-furyl) disulfide (cooked meat, rice), or 1,(E3,Z5,Z8)-undecatetraene (some plant foods) depend on more specific precursors or pathways and determine the odor of only a few food products. It is this group of highly odor-active volatile flavors that has received paramount priority in academic and industrial approaches for an innovative biotechnology of aromas.

1.3 Categories of Aromas

Aroma compounds in food may origin from microbial or plant metabolisms or, to a much lesser extent, from animal metabolism. Continuing and extending German food legislation, the aroma guideline obligatory to all EU member states (cf. Sect. 3.2) classifies all aroma compounds or mixtures as 'natural' when they have been obtained from natural sources by either physical or fermentative processes. In the US, flavors that are separated from a food source or generated during heating or processing by enzymatic activities or fermentation are all regarded as 'natural'. These regulations open opportunities for biotechnological processes.

In contrast to EU aroma terminology, synthetic compounds derived from petrochemistry or other natural chemicals which are added for flavoring purpose are considered 'artificial' in the US. In Europe, the synthetic counterparts of naturally occurring aroma compounds are classified as 'nature-identical'; only compounds without a natural prototype are 'artificial'. As a result, the status of an European 'artificial' aroma chemical will change to 'nature-identical', after it has been unambiguously identified in a natural material. During the writing of this book the commission of the EU is discussing a change of current aroma legislation to a positive list system, as has been established in the US since 1958. There, an expert panel supported by the Flavor and Extract Manufacturers association (FEMA) and the Food and Drug Administration (FDA) publishes lists of 'Generally Recognized As Safe' (GRAS)-materials. The most recent list displays about 3800 entries of extracts, essential oils, and natural and artificial (in European terms) compounds and includes maximum dosage levels anticipated for various categories of foods and beverages (Smith and Ford, 1993). If labeled accordingly, the aroma materials can be included in food formulations. All GRAS-materials have been exempted from the requirements of regular food additives. The coevolution of natural aromas and the olfactory receptors may explain, why non-naturally occurring volatiles (and many fragrance chemicals) can only imitate or vary the sensory principles of natural compounds; however, increased processing stability or other technological advantages of artificial GRAS-materials may justify their continued use.

1.4 Functions of Aromas

Food processing operations, starting with improper harvesting over extended storage to thermal treatment may cause a loss of aroma that calls for subsequent supplementation. If the losses are minor, an addition will help to enhance or round off the genuine stock of flavors. The current tendencies toward light, less, and low-everything products and to minimally processed food have created new markets for volatile flavors. Reducing fat, for instance, will inevitably reduce the flavor contents, and minimized thermal treatment (microwave) will similarly only be compensated by an appropriate addition of volatiles that match the expectations set up by our flavor memory (Hatchwell, 1994).

A carefully composed mixture of compounds may be chosen to impart specific odor attributes that will render the product unmistakable (brand name products); totally new flavor sensations have, however, only in very rare cases been accepted by the consumer for a longer period (for example, Cola beverages). Aroma compounds are also used to substitute expensive natural extracts (for example, vanillin or vanilla-type composed flavors). Finally, undesirable, natural odor notes can be masked not in order to hide spoilage, but to improve the acceptability of a nutritious product (for example, soya milk).

Following a recent market survey (Food Marketing Institute, 1993) the sensory quality of a food has been ranked higher than nutritional value, price, or safety. Going far beyond simple acceptability of food items, the early development of the chemical senses in human evolution has linked the perception of odorous compounds to emotional and, thus, behavioral responses not controlled by conscious brain functions. The ever-increasing interest of the food industry is easily explained by recognizing that volatile flavors act as chemical messengers on human beings. Likewise, the same, growing understanding has led certain consumers to express concern about highly-flavored products ('secret seducers'). However, most of the aroma-rich food items would be simply not consumed without a high flavor dosage (candies, chewing gum). No reliable data exist on total annual consumption ratios due to the above mentioned problems of quantification, but it appears safe to assume that the share in uptake of *added* volatile flavors is much smaller than the *genuine* aroma stock of foods.

1.5 Bioactivity of Aromas

Numerous studies have been conducted to shed light on the morphology and psychophysiology of human chemoreception (Burdach, 1987). Odor-evoked memories were investigated and found to be vivid, specific and relatively old. The semantic problem of describing odor impressions unequivocally was tackled in a different way by both sexes (Herz and Cupchik, 1992). Another study, using the electroencephalographic method of measuring the slow brain waves, found no bias as to sex, age, national origin, or race, when stimulating, sedating, or neutral odorants were applied (Manley, 1993). Work at the Monell Chemical Senses Center was related to the 'sick building' syndrome, exploring the effects of intermittent bursts of different odorants on task performance, mood, and perceived health (Knasko, 1993). In contrast to other reports, none of these measures was affected under experimental conditions, but test persons retrospectively reported that they thought the malodors had adverse effects. In spite of such sometimes contradictory results there appears to exist a broad agreement that volatile flavors can signal much more than just food quality (Fig.1.2).

Chemical messengers are vitally important for animals, plants, and microorganisms. Biological pest control is one attractive application. For example, 4-(4-hydroxyphenyl)-2-butanone, an impact component of raspberry flavor, has been described as an insect attractant and is used in combination with malathione as an

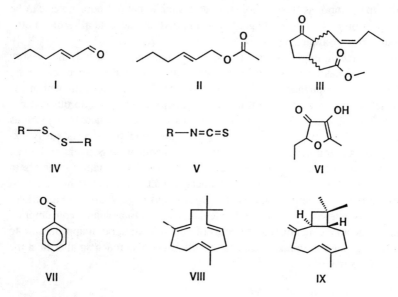

Fig. 1.2. Volatiles with multiple bioactivities. I = (E2)-Hexenal (green), II = (E2)-Hexenyl acetate (green-fruity), III = Methyl jasmonates (jasmin), IV = Dialk(en)yldisulfides (allium-like), V = Isothiocyanates (vegetable notes), VI = 2-Ethyl-4-hydroxy-5-methyl-3(2H)-furanone, VII = Benzaldehyde, VIII = Humulene, IX = β-Caryophyllene

insect bait (Perez, 1983). Volatiles can stimulate saliva flux and digestion of food, inhibit many fungi and bacteria, and may possibly even interfere with cancer formation in humans. These bioactivities above and beyond the direct sensory activity should become increasingly important for the food industry in the future. Plant and microbial experimental systems are much easier to control than humans, and the advanced methodology has allowed us to accumulate considerable knowledge in this field. For example, a checklist of floral scent chemicals acting as pollination attractants has become available (Knudsen et al., 1993). This database will assist us to evaluate volatile phytochemicals. Bioassays to estimate parameters, such as growth inhibition, and non-invasive systems to collect plant and insect volatiles have been described (Hamilton-Kemp et al., 1992; Heath and Manukian, 1992).

These achievements provide an experimental and information basis to characterize further biological properties, including therapeutic ones, of volatile compounds. Previous work has been summarized by Teranishi et al. (1993). The most recent papers are compiled in Tables 1.1 to 1.5. Comparing the mere number of references shows that animal and microbial volatiles have been researched much less than plant volatiles. Particularly well known is the defense response of wounded plant cells against microbial attack, mediated by various six and nine carbon carbonyls and alcohols (Table 1.2). The enzymatic chain starts with the oxygenation of unsaturated fatty acids by type I lipoxygenase and is similar in leaves and fruits. As observed for many other volatiles, the most potent odorants produced via the lipoxygenase pathway exhibit significant further bioactivities. A mixture of volatiles is often more potent in stimulating specific biological responses than a single constituent, (for example, Muroi et al., 1993; Byers, 1992). The phytoprotective, odorous monoterpenes are the active principles of many spices used traditionally for a 'natural' preservation of food. Very few active constituents have been investigated on the enzymatic level of formation. The sprout growth inhibition of S-carvone on potato tuber was concurrent with a strong decrease of activity of 3-hydroxy-3-methylglutaryl-CoA reductase (EC 1.1.1.34), but no direct effects on either enzyme or mRNA levels were found (Oosterhaven et al., 1993).

An increasing number of volatile flavors are reported to interact with detoxifying enzymes or carcinogenic metabolic intermediates. Caryophyllene and humulene, constituents of hop oil and other essential oils, and eugenol, the cha-

Table 1.1. Animal sources of bioactive volatiles

Source	Bioactivity	Reference
urine of rats	behaviour, aggressiveness	Garcia-Brull et al., 1993
urine of mice	accelerate puberty	Schellinck et al., 1993
secretion of minks	repellent	Epple et al., 1993
alkanals, alkanols ...	wax moth pheromone	Romel et al., 1992

Table 1.2. (Lip)oxygenase derived, bioactive plant volatiles

Compound(s)	Bioactivity	Reference
(E2)-hexenal	bactericidal	Croft et al., 1993
(E2)-hexenal/ol	bactericidal	Deng et al., 1993
(E2)-hexenal (+indol)	bactericidal	Muroi et al., 1993
Mate tea volatiles	antimicrobial	Kubo et al., 1993
hexenal/ol	decrease aphid fecundity	Hildebrand et al., 1993
(E2)-hexenal	fruit fly repellent	Scarpati et al., 1993
C6 to C10-enals	elicite defense response	Zeringue, 1992
C6 enoyl esters ...	wasp orientation	Whitman and Eller, 1992
fruit flavor compunds	antifungal	Vaughn et al., 1993
(E2)/(Z3)-nonenal ...	antifungal	Vaughn and Gardner, 1993
C6 and C9 compounds	antifungal	Hamilton-Kemp et al., 1992
'leaf odors'	aphid attractant	Dwumfour, 1992
'methyl jasmonate'	inhibits ester formation	Olias et al., 1992
jasmonates	plant movement	Weiler et al., 1993
jasmonate isomers	inhibit plant growth ...	Koda et al., 1992

Table 1.3. Bioactive volatile isoprenoids from plants

Compound(s)	Bioactivity	Reference
Majorana essential oils	antifungal	Shimoni et al., 1993
limonene, geraniol	antifungal	Oh et al., 1993
'monoterpenes'	antimicrobial	Himejima et al., 1992
β-pinene...	inhibits plant pathogens	Donovan et al., 1993
δ-cadinene,β-caryophyllene	antibacterial	Muroi and Kubo, 1993
perillaldehyde	inhibits fungi and bacteria	Kang et al., 1992
nerol, geraniol ...	inhibit algal growth	Ikawa et al., 1992
α-pinene, 3-carene...	attract insects	Byers, 1992
'flower volatiles'	attract moths	Zhu et al., 1993
α-terpineol, geraniol ...	phytotoxic	Vaughn and Spencer, 1993
S-carvone	inhibits sprouting	Oosterhaven et al., 1993
sesquiterpenes	'anti-carcinogenic'	Zheng et al., 1992

Table 1.4. Bioactive plant volatiles from different chemical classes

Compound(s)	Bioactivity	Reference
various volatiles	stimulate spore germination	French, 1992
allyl isothiocyanate	inhibits yeast growth	Tokuoka et al., 1992
2-butanone	antifungal	Kumar, 1993
furfural	nematicide	Rodriguez-Kabana et al., 1993
acetaldehyde, ethanol	activate fungi	Cruickshank & Wade, 1992
cinnamaldehyde, thymol...	inhibit potato tuber sprouting	Vaughn & Spencer, 1991
'odor mixtures'	foraging behaviour of honeybee	Masson et al., 1993
dimethyl disulfide	attracts flies	Green et al., 1993
'fig volatiles'	attract wasps	Ware et al., 1993
isothiocyanates	attract seed weevil	Bartlet et al., 1993
Allium volatiles	affect insect behaviour	Nowbahari & Thibout, 1992
2-phenyl ethanol	affect seed germination	Matsuoka et al., 1993
pentanol, maltol, vanillin	affect grain pest insects	Phillips et al., 1993
furanones	insect pheromone	Farine et al., 1993

Table 1.5. Bioactive microbial volatiles

Source	Bioactivity	Reference
Rhizopus arrhizus	inhibit *Aspergillus flavus*	Lanciotti & Guerzoni, 1993
C4 to C9 ketones	inhibit plant growth	Bradow, 1993
Bacillus subtilis	fungal growth	Fiddaman & Rossall, 1993
Ceratocystis fagacearum	attract insect vectors	Lin & Phelan, 1992
i-pentanol, butanol ...	attract palm weevil	Nagnan et al., 1992
dialkyl disulfides	affect plant parasites	Thibout et al., 1993
2-ethyl-4-hydroxy-5-methyl-3(2H)-furanone	'anti-carcinogenic'	Nagahara et al., 1992

racter impact component of clove oil, were inducers of glutathione S-transferase in mouse liver (Zheng et al., 1992). Limonene, carvone, and various volatile sulfides are thought to inhibit cytochrom P450 dependent monooxygenases, thereby preventing the activation of procarcinogens. MAILLARD volatiles, such as furanones and various reductones, due to their autioxidative power may, like other antioxidants, scavenge cancer inducing free radicals (Elizalde et al., 1992). In conclusion, the various bioactivities of volatile flavors have a great interaction with current trends in food technology: the consumers demands for healthy food, minimally processed food, and 'clean' labeled food can be met by an increased application of natural active principles. These compounds, as intermediates of the metabolisms of living cells, could all be delivered more constantly and reliably by biotechnological processes.

2 The Roots: Empirical Food Biotechnologies and Formation of Aroma Compounds

Our ancestors observed that milk, wet cereal flour, fruit juices, and raw meat, when incubated for some time, underwent changes that led to more stable products (for example, Michel et al., 1992). The resulting foods, altered in texture, color, acidity, gas content, turbidity, and flavor apparently had no adverse effects on human well-being, if consumed in moderation. These very roots of modern biotechnology have evolved from artisan levels into major industries. The present output of the traditional biotechnologies far exceeds the new fermentation products in both volume and product value. According to recent year books the annual biotechnology of antibiotics is worth about 50 bio. US$, while wine and beer production amounted to an estimated 300 bio. US$. A large number of textbooks, encyclopedias and original papers have discussed all the facets of the traditional, fermented foods. This chapter will not recapitulate earlier reviews, but, subdivided under commodity categories, discuss the most recent aroma aspects. The dominating topics will be:

- of the aroma profile of existing fermented products, and
- the possible transfer of existing knowledge and proven technology to novel processes.

2.1 Characterization of Yeasts

For many years *Saccharomyces cerevisiae* has been the workhorse of microbiologists and biochemists, and is now the best-known eukaryotic organism. With this background it is embarrassing to follow the contradictory discussions of elementary properties, starting with taxonomy: While some authors call the brewing yeasts *S. pastorianus* (replacing the obsolete *S. carlsbergensis*), others consider all baker's yeasts, brewer's yeasts, wine, sake, and champagne yeasts to be strains of *S. cerevisiae* (Beneke and Stevenson, 1987). The formation of ethanol is said to be product-inhibited at levels of 12 to 15% v/v. A commercial dry yeast, supplemented with yeast extract, was able to produce as much as 21.5% v/v at high pitching rates (Thomas and Ingledew, 1992). The same level of ethanol was reached in another study by supplementations of soybean or groundnut (Ezeogu and Emeruwa, 1993). Differences in maximum ethanol production of enological yeast strains were explained by a loss of the hexose transport activity during the fermentation

(Mauricio and Salmon, 1992). A *S. cerevisiae* strain isolated from spent sulfite liquor fermentation utilized galactose. This study compared the carbohydrate enzymology of several *S. cerevisiae* strains (Linden et al., 1992).

Amino acids are the precursors of common fermentation alcohols and carbonyls, such as the methylbutanols, 2-phenyl ethanol, and their esters. The pool of amino acids is filled by intermediates of glycolysis and tricarboxylic acid pathway. Recent work has confirmed that the concentration of amino acids derived from glycolysis decreases in respiring cells (Martinez-Force and Benitez, 1992). The opposite changes take place when shifting from respiratory to fermentative conditions. In continuous culture and at low dilution rates, however, the concentration of the amino acids, whatever their precursor, increases, indicating that growth rate and phase has a more significant impact on amino acid metabolism. This should be considered when transferring batch cultivations to continuous processes.

The future biochemical and genetic differentiation of yeast will be improved by simple and highly discriminating methods, such as pulsed field gel electrophoresis of chromosomal nucleic acids (Frezier and Dubourdieu, 1992; Kelly et al., 1993). Less suitable for distinguishing strains, but useful for certain applications, are DNA finger printing and SDS-polyacrylamide gel electrophoresis of secretory proteins (Kunze et al; 1993).

2.2 Aroma Development in Beer

As for all products of empirical biotechnologies, the volatile composition of beer results from the complex interactions of the chemistry of the raw materials, the processing steps, microbial activities and maturation. Recent technological developments in brewing are not directed toward aroma improvement, but at overcoming long fermentation and aging times. Novel processes, for example using immobilized cells, are frequently associated with a significant change of the aroma profile. To compensate for flavor losses, a two stage fermentation combining free cell fermentation of the wort with immobilized cell treatment of the young beer has been reported (Russell and Stewart, 1992). Another immobilized column reactor was continuously operated, and not a gain in flavor, but savings on space and investment were emphasized. Immobilized yeast has also been suggested for operating under respiratory conditions for the generation of ethanol-reduced beer. In a maltose medium, the production of esters by immobilized yeast was poor, but was improved to free cell levels after addition of wort treated with glucoamylase (Shindo et al., 1992). Immobilization allows a higher volumetric cell mass and facilitates continuous operation. These advantages are opposed by heterogeneous cell distribution and physiology, diffusional limitations, risk of contamination, and extended build-up times of the process. The low metabolic activity of entrapped yeast is particularly reflected by a decrease of the energy (here oxygen) dependent formation of volatile esters. For example, the concentration of 2/3-methybutyl acetate was 600 times lower in an alginate-entrapped yeast as compared to the free cell fermentation (Masschelein, 1989). Similarly, a rise of the fermentation temperature to accelerate the production did shift the original aroma profile; the concentrations of higher alcohols, acetohydroxy acids, and vicinal diketones, increased, while the esters behaved irregularly (Masschelein, 1989).

Flash fermentation under reduced pressure (Sakaguchi et al., 1990) and under CO_2 pressure also affected the production of volatiles (Renger et al., 1992). The observed reduction in the formation of esters and higher alcohols was indirectly associated with CO_2 partial pressure and caused by reduced yeast growth and uptake of amino acids. The biochemistry of this effect is not understood; however, at least the size, geometry, and agitation speed of brewing vessels can be correlated with the concentration of CO_2 in different zones, and flavor formation can to a certain extent be controlled by an appropriate technical design (Masschelein, 1989), and cultivation conditions (Kuriyama et al., 1993). Carbohydrate and amino acid metabolism of yeasts with a different CO_2 inhibition was improved by the addition of protein-based chemicals (Kruger et al., 1992). An increase in higher alcohols and acetate esters, and a decrease in acetaldehyde were observed, whereas SO_2 levels remained uneffected. The sensitivity to CO_2 inhibition was strain-dependent for all parameters.

Yeasts also affected the volatile composition of beer by producing hydrogen sulfide (Walker and Simpson, 1993). Methane thiol was a precursor of methyl thioacetate in both ale and lager strains (Fig. 2.1). A second, indirect effect of the sulfur metabolism of yeast is based on sulfite production (Kaneda et al., 1992). Various technical measures favored sulfite generation and, thereby, a better stability of beer flavor. These included a reduction of dissolved oxygen in the pitching wort, a higher pitching rate of the yeast, and better wort filtration. The flavor character of the treated, fresh beer was claimed not to have changed by the altered processing conditions.

Rapid fermentation technology and immobilization of yeast have both been reported to intensify the problem of formation of the 'butterscotch' flavor compounds, diacetyl and 2,3-pentandione. Under less anaerobic conditions cells grow well and overproduce 2-acetolactate and 2-aceto-2-hydroxy-butanoate, intermediates of the pathways to valine and isoleucine. The acids are excreted into the wort and decarboxylate spontaneously to yield the above vicinal diketones. To remedy the resulting flavor defect the use of genetically engineered yeast has been described (cf. Sect. 8.2.3). As problems of obtaining approval and consumer acceptance are to be expected, other modifications of the brewing process are being pursued further. The production of 2-acetolactate by immobilized yeast can, for example,

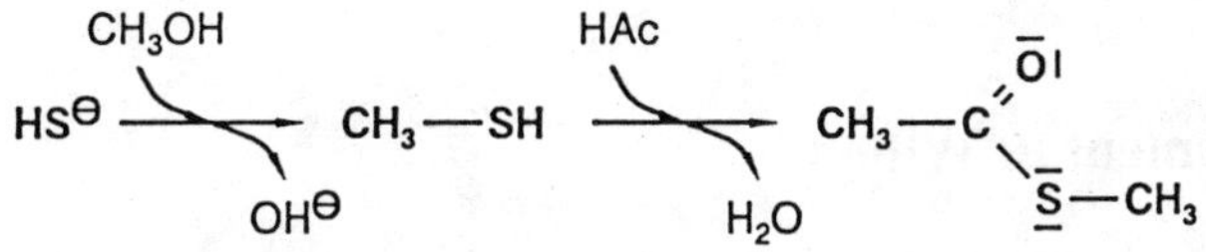

Fig. 2.1. Yeast induced formation of thio compounds in ale and lager beer (Walker and Simpson, 1993)

be reduced by changing the immobilization matrix (Shindo et al., 1993). The effectiveness of double-layered gel fibers in reducing precursor acid formation was attributed to a reduced pO_2 inside the gel fibers. This, in turn, slowed down the formation of the undesired odorous compounds. An alternative solution can be seen in a concerted use of 2,3-butandione and pentandione converting enzymatic activities (Heidlas and Tressl, 1990). In addition to alcohol dehydrogenase, three further enzymes were found to reduce the diketones: (S)- and (R)-diacetyl reductase, and (R)-2,3-butandiol dehydrogenase. Both diacetyl reductases can be distinguished by classical parameters including the different substrate and product specificities. The resulting acyloins could then be further reduced to (R,S)- or (R,R)-2,3-butandiol according to the stereochemistry of the precursor (Fig. 2.2). As soon as technical parameters, such as pO_2 or temperature, can be successfully correlated with the activity of the reducing enzymes, the conversion of the potent diketones into odorless (at actual concentrations) products will be greatly facilitated. Many data on the stereoisomer 2,3-butandiols were recently presented for wine (Hupf and Schmid, 1994).

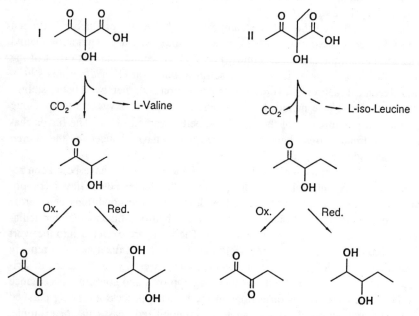

Fig. 2.2. 2-Aceto-2-hydroxypropanoate (I) and -butanoate (II) are precursors of vicinal diketones and diols in beer

2.3 Aroma Development in Wine

Winemaking, as a part of man's cultural history, is still regarded as a combination of the viticulturist's art and sophisticated technology. The 'biological-technological' (biotechnological) 'sequence of wine' has been formulated to explain the stepwise generation and changes of volatiles up to the finished wine (Drawert, 1974). Using a purge/extraction technique the time course of volatile formation was monitored during the entire fermentation, and maximum levels were found on the third day (Stashenko, 1992). As reflected by the labeling regulations, there is general consent that both grape cultivar and soil have a major impact on aroma type and quality. The TNO compilation lists more than 450 volatiles for grape, and about 650 compounds for wine. Though the mere number of compounds does not allow extensive conclusions on aroma quality, the data suggest a significant contribution from microbiology (yeast, *Botrytis cinerea* (noble rot), *Leuconostoc oenos*).

2.3.1 Yeast and Wine Aroma

Yeast cells generate volatiles during growth, liberate volatiles from nonvolatile grape precursors, and may catabolize off-notes, such as thio compounds or vinylphenols (Ribereau-Gayon, 1993). Among the often cited impact volatiles of wine are: 3-methylbutanol, 3-methylbutyl acetate, and the ethyl esters of acetic hexanoic and octanoic acid. Similar short to medium chain alkanols, carboxylic acids, and corresponding esters dominate the aroma of sake. The formation of these compounds clearly depends on yeast enzymes, and is affected by pO_2, must turbidity, contact to peel, and maturation. However, different wines with a similar content of higher alcohols and esters may smell differently, indicating a specific role of minor and trace impact components. According to a recent study (Fischer, 1994) most of the volatile contributors to the aroma of a *Chardonnay* wine are related to yeast metabolism (Table 2.1). The origin of heterocyclic constituents, such as furaneol, is more difficult to assess. A liberation from glycosidic and other precursors has been made likely (cf. Sect. 7.2.1 and below). Other grape varieties produce a group of oxygenated, odorous monoterpenes that is transferred into the wine. The de novo formation or bioconversion of monoterpenes has been a subject of controversy, but recent, thorough studies have demonstrated the very limited potential of regular *S. cerevisiae* in this respect (Hock et al., 1984). Obviously the biosynthesis of monoterpenes is confined to (wild) yeast strains that are rather suppressed in modern winemaking.

When must of the Spanish grape cultivar *Monastrell* was fermented with different strains of *S. cerevisiae,* major differences in the ester content of the wines were observed (Mateo et al., 1992). Statistic treatment of the data led to the conclusion that highly correlated volatiles were not derived from the same metabolic pathway. It was assumed that the dynamic interconversion of compounds may explain a correlation of pairs or groups of volatiles. More obvious were correlations of *n*-propanol and hydrogen sulfide formation (Giudici et al, 1993). It was concluded that the production of *n*-propanol was related to methionine and threonine metabolism. The emanation of hydrogen sulfide during fermentation was followed by Thomas et al. (1993). Two maxima occurred independently; one was associated with the yeast strain used, the other one with the medium. Elemental

Table 2.1. Aroma dilution values of a *Chardonnay-Semillion* wine according to gas chromatography-olfactometry (Fischer, 1994)

Compound	Dilution Factor $[10^X]$
	X
❑ Carbonyls	
3-heptanone	1
β-damascenone	5
4-hydroxy-3-methyloctanoic acid lactone	1
4-hydroxy-2,5-dimethyl-2(3H)-furanone	1
❑ Alcohols	
3-methyl-1-butanol	4
2-phenyl ethanol	5
❑ Acids	
acetic	2
butanoic	1
3-methylbutanoic	2
pentanoic	1
hexanoic	1
octanoic	1
❑ Esters	
3-methylbutyl acetate	4
hexyl acetate	1
ethyl propanoate	2
ethyl 2-methylpropanoate	1
ethyl butanoate	4
ethyl 2-methylbutanoate	2
ethyl 3-methylbutanoate	1
ethyl hexanoate	4
ethyl octanoate	3
ethyl dodecanoate	1
methyl hexanoate	1
methyl octanoate	1
❑ Miscellaneous	
3-methyl thio-1-propanol	1
3-methyl thiopropanoic acid	1
2-methoxy-4-vinylphenol	2
2-methoxy-4-allylphenol (eugenol)	5
4-hydroxy-3-methoxybenzaldehyde (vanillin)	1

sulfur was added to the medium, but no relation to the amount of hydrogen sulfide was found. Factors affecting the production of volatile sulfur compounds in wine and beer have been systematically evaluated in a recent study (Park, 1993). When a *Sauvignon blanc* juice was fermented using different yeast strains, different amounts of hydrogen sulfide were found. In a glutathione supplemented medium, hydrogen sulfide, sulfur dioxide, carbonyl sulfide, dimethyl sulfide, and ethane thiol were found. With the exception of hydrogen sulfide, these sulfur volatiles completely disappeared during fermentation.

The over-all importance of the yeast becomes particularly apparent, if the other variables of the process are kept constant. A Spanish must was fermented using 14 isolates from three wineries (Longo et al., 1992). Variations in the resulting concentrations of acetaldehyde, 2-methylpropanol, pentyl alcohols, and ethyl acetate were significant in laboratory scale fermentations. The most comprehensive study in the field was conducted on a synthetic medium. Romano et al. (1992) compared 69 apiculate wine yeasts for their ability to produce higher alcohols and acetic acid. These and other authors (Holloway and Subden, 1991; Ranadive and Pai, 1991; Mateo et al., 1991) have emphasized the strain and species dependent formation of volatiles. Yeast of lower fermentative power, such as *Candida*, *Kloeckera apiculata*, or *Hansenula anomala* exceeded commercial *Saccharomyces* strains in terms of volatiles formation. Not all these studies have used wine-related media which somewhat restricts a direct transfer of the data to technical scale wine production. However, when wines from three *Chardonnay* musts were fermented by two different strains of *S. cerevisiae*, one group of products could be directly correlated to the yeast: 3-methylbutyl acetate, ethyl dodecanoate, and the sum of esters (Delteil and Jarry, 1991). Among the constituents related to both yeast and must were acetic acid, 2-methylpropanol, glycerol, hexanol, and various ethyl esters. As a result, the aroma of the different wines could be clearly distinguished.

2.3.2 Yeast Enzymes

Present knowledge of the particular enzymatic activities of the yeast involved in aroma formation is limited. The formation of esters and higher alcohols was shown to be more dependent on the protein substrate than on yeast protease; again, for the same substrate the yeast's protease activity becomes important (Rosi et al., 1990). Acetyl transferase (EC 2.3.1.84) activity developed early during fermentation and coincided with the increase of ethyl, 3-methylbutyl, and hexyl acetate esters (Mauricio et al., 1993). After ten days a concurrent increase in esterase activity and hydrolysis of esters was observed. The substrate specifity of the esterases deviated within extreme limits (Rosi et al., 1990). Supplementing fermenting yeasts with fatty acids or triacylglycerol mixtures resulted in an altered composition of the cellular lipids (Rosi and Bertuccioli, 1992). In turn, the excretion into the medium of esters, higher alcohols, and medium chain fatty acids was significantly changed. Higher levels of unsaturated C18 fatty acids led to increased formation of higher alcohols, and an increase of C10 to C16:1 fatty acids correlated with the formation of medium chain fatty acids and esters. Some speculations on the origins of these esters from β-oxidation, fatty acid synthesis, and from 2-keto acids are attractive, but have not yet been confirmed experimentally. After the identification of glycosidically bound monoterpenes in grape must (cf. Sect. 7.2.1), another class of hydrolases has received considerable interest: Glycosidases of *S. cerevisiae* cleave β-glycosidic bonds and are little inhibited by free glucose. The use of yeasts with high activities of β-glycosidase may improve the aroma of white wine, but is less favorable for red wines because of the simultaneous attack on anthocyans and flavonol pigments.

Vinyl and ethylphenols are among the character impact constituents of red and white wines (cf. Table 2.1). They result from the nonoxidative decarboxylation of phenylpropanoid precursor acids and a subsequent vinylphenol reductase activity (Chatonnet et al., 1993). The intracellular cinnamate decarboxylase was constitutivly formed and only expressed during the phase of ethanol production. The enzyme showed inhibition toward procyanidins and catechins which explained why red and rose wines contained lower levels of the degraded phenols in spite of containing more precursors. The same enzyme from wild yeasts was not inhibited by polyphenols which resulted in high amounts of ethylphenols in red wines with phenolic off-odor (Chatonnet et el., 1992). Electrophoretic karyotyping was suggested for selecting yeast strains that do not express this less desirable trait (Grando et al., 1993).

2.3.3 Noble Rot Infected Grape

Famous sweet wines, such as *Tokay Aszu*, *Sauternes*, or *Trockenbeerenauslese* are made of fully ripe grapes infected by *Botrytis cinerea*. Enzymes of *B. cinerea* are thought to be involved in the formation of sotolone (4,5-dimethyl-3-hydroxy-2(5*H*)-furanone), an impact component with caramel/walnut notes (Fig. 2.3). Similar to the formation of diketones/diols in beer (Fig. 2.2) an oxoacid derived from an amino acid could serve as precursor. Chain elongation by an unknown aldolase could yield the immediate, multifunctional precursor of the keto tautomer of sotolone. Stereochemical data did not support the hypothesis of enzymatic formation for all samples examined, but speculations are misplaced as long as the correct pathway remains unknown (Guichard et al., 1992). The same authors suggest that flor-sherry yeast may produce sotolone in a suitable chemical environment. The structurally related solerone (5-acetyldihydro-2(3*H*)-furanone, 'bottle' note) is widely spread in wine. The precursor acid may be derived from glutamic acid via α-ketoglutaric acid. The head to head condensation of the two carbon aldehyde would again lead to the 1,4-diol configuration required for 4-ring formation (Fig. 2.3). As with sotolone, the optical purity of solerones isolated from wine samples does not require an enzymatic explanation, but enolization of the formed product during maturation, or subsequent enantioselective bioconversion could be responsible for the observed racemization (Guichard et al., 1992). Unlike yeast, *B. cinerea*, as an ascomycete, has been demonstrated to convert grape monoterpenols and citral (Brunerie et al., 1988). A series of redox reactions and, remarkably, also ω-hydroxylations were reported (Fig. 2.4). The quantitative effect of these bioconversions on the aroma of a botrytized wine has not been elucidated so far.

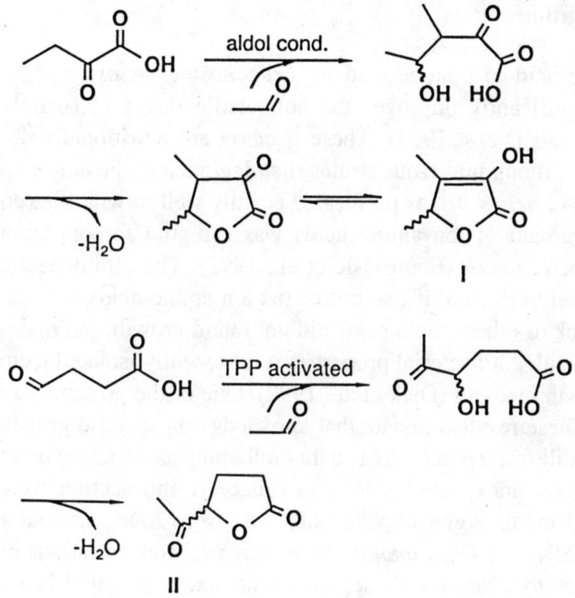

Fig. 2.3. Proposed biogenesis of sotolone (I) and solerone (II), impact components of some wines and sherry

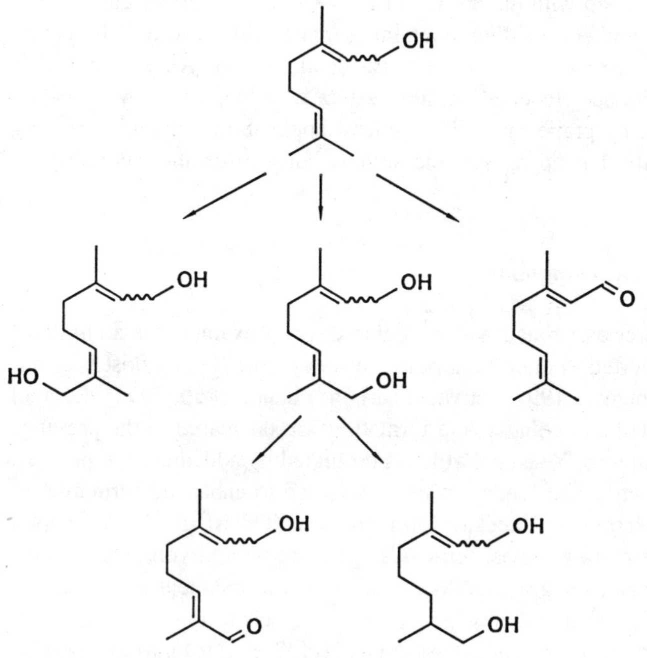

Fig. 2.4. Conversion of monoterpenols by *Botrytis cinerea* (Brunerie et al., 1988)

2.3.4 Malolactic Fermentation

The conversion of L-malic acid to L-lactic acid by *Leuconostoc oenos* or other *Lactobacilli/Pediococci* significantly improves the nonvolatile flavor of (mainly white) wines (van Vuuren and Dicks, 1993). These bacteria are nutritionally demanding, and, for example, among numerous strains that degraded malic acid in a synthetic medium only two *L. oenos* strains performed equally well in wine (Pardo and Zuniga, 1992). A minimum of ten amino acids was essential for optimum growth of various strains of *L. oenos* (Fourcassie et al., 1992). The rate of lactic acid production was adversely affected, if one out of the ten amino acids was deficient in the medium. A lack of other amino acids did not retard growth, but malolactic conversion. Proteins with antibacterial properties were recently isolated from *S. cerevisiae* cells and growth medium (Dick et al., 1992). One of the proteins had lysozyme characteristics. These results indicate that knowledge on specific growth factors and interactions is still fragmentary, as is data on the impact of *L. oenos* of the spectrum of volatiles. Not unexpectedly, levels of diacetyl and acetoin were increased. On wine-related media some aliphatic alcohols were found as major volatiles (le Roux et al., 1989). In a *Chardonnay* must, increased concentrations or formation of acetic acid, methylpropanol, 4-methyl-3-pentenoic acid, ethyl hexanoate, hexyl acetate, and farnesol were detected after malolactic fermentation by coupled gas chromatography-mass spectrometry (Avedovech et al., 1992). These authors also reported phthalic acid and other constituents as not found in control wines without malolactic fermentation. In summary, the origins and impact of various volatiles associated with the growth of *L. oenos* awaits further elucidation. Deacidification of a synthetic medium containing malic acid could also be achieved using *Schizosaccharomyces pombe* (Sousa et al., 1993). An external loop reactor was used to induce flocculation, and results were improved by replacing the synthetic medium by grape must. The general applicability of this intriguing approach will not only depend on suitable strains, but also on the technological implications.

2.3.5 Technological Developments

Technological measures associated with microbiology, for example the addition of biosorbents, were reported to alter the aroma profile of wine (yeast ghost preparations: Minarik and Jungova, 1992; activated carbon: Fukuda et al., 1991). Ethanol formation was activated and volatile acid formation was decreased in the presence of yeast ghosts or cellulose. Yeast growth was promoted by added carbon powder, and the concurrent increase of leucine consumption led to enhanced formation of 3-methylbutanol. 3-Methylbutyl acetate formation was highest in carbon supplemented mash without stirring. Nonspecific adsorption of neutral volatiles onto the biosorbents can be taken for granted. As a result, such a treatment may improve the volatile flavor of sake or other wines, because co-adsorption of sensory impacts is prevented in the presence of excess ethanol (cf. Sect. 2.9.1). Many workers in the field still focus on primary process parameters, such as cell growth or ethanol formation, while flavor changes remain open (Parascandola et al., 1992). More polar immobilization matrices, such as ion exchange resins, were suggested to accelerate cider fermentation (O'Reilly and Scott, 1993). Similar effects based on decreased pCO_2 could also be beneficial to regular grape must fermentation. In fact, the empirical process creates large interfaces between must and biosorbents (grape epidermal tissue, stalks, yeast flocs) which are known to influence the sensory attributes of the finished product.

Cryophilic strain of S. cerevisiae were selected and found to produce less ethanol at intermediate temperatures of 22 °C and 30 °C, respectively (Kishimoto et al., 1993). In another study a cryophilic strain was immobilized on an inorganic support (Bakoyianis et al., 1993). Ethanol production at 5 °C was equal to a regular fermentation at 22 to 25 °C. This was maintained during a 75-day continuously operated fermentation and assigned to the immobilizing mineral matrix. Total and volatile acidity, and the concentrations of main volatiles decreased, while ethyl acetate concentrations increased as compared to freely suspended cells. Techniques of yeast cell immobilization are as manifold as the required technical alterations of the process as a whole. While the idea of an immobilized catalyst is attractive in operational terms, many biochemical consequences have not been discovered yet. Restricted external mass transfer can lead to concentration gradients in the surroundings of entrapped cells, which will lower the toxicity of ethanol on its producer cell. The changed water activity inside the gel and the improved availability of reduced nucleotide coenzymes may account for increased levels of certain volatiles. The confinement in gels may also increase the copy number of extrachromosomal nucleic acids. The network character of factors involved in the metabolism of yeast volatiles calls for more detailed flavor studies to evaluate the biochemical basis prior to the upscaling of a process.

2.4 Yeast Cells as a Source of Aroma

Yeast biomass as such can be processed into flavoring materials. Developed from spent brewer's yeast, modern processes yield an annual output of about 35 000 metric tons (Nagodawithana, 1992). There are three major manufacturing practices based on endogenous hydrolytic and accessory enzymes (autolysis), the addition of common salt (plasmolysis), and hot hydrochloric acid (hydrolysis). Large amounts of yeast wastes are available rendering the process appealing to the food industry. The dominating RNA fraction of yeast is the precursor pool for the generation of 5'-nucleotides that are useful taste enhancers in sour and salty foods, particularly in the presence of glutamate. Constituents of the volatile fraction impart all kinds of odor impressions, frequently meaty, sulfury, and, of course, yeast-like notes. Recently conducted comprehensive studies identified hundreds of volatile constituents (Werkhoff et al., 1991; Ames and Elmore, 1992). The volatiles originated from pathways of the MAILLARD reaction, and the contribution of thiamin to the formation of sulfur compounds was emphasized. Among the odorants identified were sulfur-substituted oxygen and mixed heterocycles, bicyclic sulfur compounds, and linear and cyclic polysulfides. The biochemistry of autolysis starts with an autocatalytic degradation of yeast proteins in which an endoprotease was characterized as an important contributor (Slaughter and Nomura, 1992). The majority of biochemical as well as chemical routes is still a matter of speculation.

Fermentation of molasses by yeasts yielded a salty seasoning (Ito and Toeda, 1993). The same bioprocess can be conducted using *Zymomonas mobilis*, an ethanol producing bacterium (Iida et al., 1992). Its biochemical uniqueness has aroused a lot of scientific interest in recent years. Immobilization helps to overcome the poor tolerance to salts in molasses or related waste streams. In a continuous process, 60 g ethanol$\times$L^{-1}h^{-1} were produced. A selected *Zymomonas* strain coimmobilized with invertase, allowed theoretical yields of ethanol to be produced in sucrose based media (Kirk and Doelle, 1993). No data on other volatiles have been reported, but, due to its prokaryotic nature, a significant aroma impact is arguable.

2.5 Aromas of Fermented Milk

Products resulting from the coagulation of milk represent another important sector of traditional food biotechnology. The starter cultures of dairy technology are mainly procaryotes, namely *Lactococci, Lactobacilli, Leuconostocs, Bifidobacteria, Propionibacteria, Streptococci*, and *Brevibacterium linens* (Hammes, 1990). In a manufacturing plant, consistent and rapid acid production is still the top criterion for the selection of starter culture strains; less importance is attached to digestibility, nutritional value, texture or the inhibition of pathogenic contaminants (Kim, 1993; Jeppesen and Huss, 1993, Hammes and Tichaczek, 1994). The ongoing discussion of 'probiotic' effects of fermented milk products includes lowering blood cholesterol concentration, stimulation of the immuno system (Marshall, 1993), and even biotherapeutic applications (Demacias et al., 1993). Some of these views have been strongly challenged.

Much more obvious is the microbially induced gain of flavor. As the potential of acidification is by no means correlated with the formation of volatiles, starter cultures have never been selected with respect to the aroma criterion (Pfleger, 1992). Looking at the sensory profiles of fermented milk products, surprisingly little attempts have been made to improve control and improve the microbial potential. Analytical separation problems caused by the fatty matrix have been solved by dynamic headspace enrichment (Collin et al, 1993; Yang and Min, 1993) or high vacuum distillation techniques (Preininger et al., 1994). Neither acetaldehyde for yoghurt (Kneifel et al., 1992; Laye et al., 1993) nor the couple diacetyl/acetoin nor other impact components can solely determine the volatile flavor of a milk product. This has been particularly well documented for the various types of cheese, where short chain fatty acids, 2-ketones, lactones, sulfur and nitrogen compounds, heterocycles and others bring about the different aroma profiles (Hammond, 1989; Imhof and Bosset, 1994). Using liquid CO_2 a series of nonvolatile, low molecular weight flavors including N-acyl amino acids were isolated from a French cheese (Roudotalgaron et al., 1993). Some of the compounds were bitter, and N-propanoyl methionine had a cheese flavor. This recently accumulated, detailed knowledge should provide the basis for overcoming the usual, narrow view of diacetyl/acetoin as the only indicators of microbial aroma formation (Bassit et al., 1993).

The flavor of milk products can be improved by simply adding β-D-galactosidase (EC 3.2.1.23) to hydrolyze lactose into more sweet and better soluble carbohydrates. The EMC (enzyme modified cheese) processes represents an established multienzyme treatment of milk fractions for the generation of aroma enhanced semi-finished products. Suitable selection and dosage of lipases/proteases accelerate the aroma formation by imitating the traditional maturation. The proteolytic enzymes of starter bacteria have been described as an arrangement of endo and aminoproteases, and tri and dipeptidases (Visser, 1993). Peptidase and esterase activities of cheese related bacteria were cell-wall associated (Ezzat et al., 1993). ^{14}C-labeled casein was used to measure proteolytic activity. Esterases of *Propionibacterium* possessed different substrate specificities (Dupuis and Boyaval, 1993). Biochemical characteristics of D-lactate dehydrogenase and oxalacetate decarboxylase, both involved in the formation of aroma carbonyls, were investigated (Fitzgerald et al., 1992; Hugenholtz et al., 1993). The partially unexpected results (the *Lactococcus* decarboxylase did not contain biotin) underscore the necessity of more enzymological studies in the field.

Complementary activities of various species during cheese ripening have been reported repeatedly (Olson, 1990; Requena et al., 1993). This agrees with observations on the sensory improvements upon adding active biocatalyst in a traditional process (Table 2.2). Especially low fat (which is low aroma) cheeses appear to benefit from these additions, but there is a permanent concern about bitter off-flavors occurring (Muir et al., 1992; Trepanier et al., 1992). Milk, as a food of animal origin, introduces less volatile flavor into the fermentation process than a fruit or vegetable material. As a result, the microbial contribution must be estima-

Table 2.2. Biocatalytically active additives to improve the aroma of cheese

Supplementation	Reference
Enzymes from A. oryzae	Fernandez-Garcia et al., 1993a
Neutrase/Palatase	Fernandez-Garcia et al., 1993b
Peptidase/Protease	Muir et al., 1992
Heat-shocked Lactobacillus	Trepanier et al., 1992
Micrococcus	Lee et al., 1992a

ted as being even higher than for fermented beverages, for example. Cheeses made with different single starter strains received different organoleptic scores (Law et al., 1992). The results indicated that the strains catabolized milk constituents in different ways. The preferred cheese contained a threefold amount of free amino acids.

Research along these lines has traced two major obstacles to generating aroma in milk substrates: the limited biosynthetic abilities of lactic acid bacteria, and the insufficient concentrations of precursor compounds. The supplementation of biocatalysts as described above, or of substrates, such as citrate or threonine (Marshall, 1987) can lead to improvements on an empirical basis. Recently, patents were granted (Okonogi et al., 1992; Verhue et al., 1992) for the biotechnological production of diacetyl demonstrating that a successful industrial scale process can be developed without a complete picture of all the biochemical details (Fig. 2.5). Another volatile target compound was propanoic acid (Lewis and Yang, 1992; Babuchowski at al., 1993). Batch cultures of *Propionibacterium* were used to examine the conversion of different carbohydrates and lactate. Hydrolyzed corn starch and lactate were preferred, and acid concentrations of more than 30 g×L^{-1} were reached. Acetaldehyde was formed in cow's milk and soy extract by a *Bifidobacterium*, but maximum concentrations remained in the lower mg range (Murti et al., 1992).

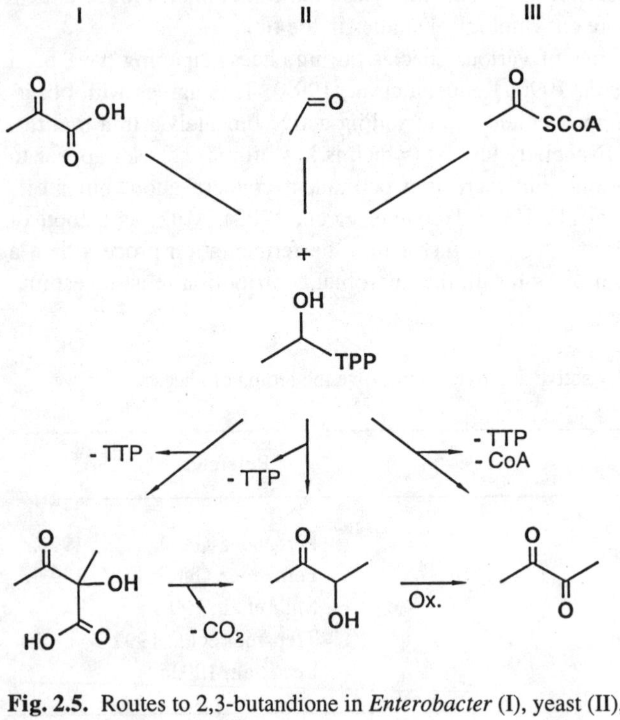

Fig. 2.5. Routes to 2,3-butandione in *Enterobacter* (I), yeast (II), and lactic bacteria (III)

Table 2.3. Immobilized starter cultures

Species immobilized	Matrix	Reference
L. lactis	Alginate	Cachon and Divies, 1993
L. lactis	Gelatin	Hyndman et al., 1993
L. lactis	Chitosan	Groboillot et al., 1993
Lactobacillus sp.	Pectate	Richter et al., 1992
L. lactis	Alginate	Champagne et al., 1992
L. rhamnosus	Various	Goncalves et al., 1992
Propionib. shermanii	Various	Begin et al., 1992

Immobilized lactics continue to attract interest (Table 2.3). The preparation is often associated with a loss of metabolic activity, but sometimes the cells recover after being recycled into the process. Diffusional limitations can be reduced by using hollow spheres with extremely thin walls or by adsorption onto suitable supports. Adsorbed cells, however, tend to gradually desorb or may lose activity upon collision with support particles. Products of immobilized dairy strains were mainly the naming acids and diacetyl. Attention was paid to aspects of preparation, stability, and ease of operation. For example, ethanol treatment or heating of the beads maintained acidification activity of *Lactococcus lactis* (Champagne et al., 1992). Volatile flavors were not mentioned by most of these authors.

Yeasts, such as *Kluyveromyces*, *Debaromyces*, *Candida* or *Trichosporon* strains, and prokaryotics, such as *Brevibacterium* have been identified in many traditionally manufactured milk products (Kaminarides et al., 1992; Besancon et al., 1992). These microorganisms synthesize or assimilate volatile nitrogen and sulfur compounds and, thereby, modify the sensory character of the products. Other strains are safely used for the fermentation of vegetables (Meurer and Gierschner, 1992), particularly soya beans (Murti et al., 1993a; Iwasaki et al., 1993). Looking at the universality of biochemistry, tradition may be a major reason for not using these microorganisms in dairy starters; their incidental occurrence suggests that, upon sufficient adaptation, these strains survive in the substrate milk and may offer the potential of creating novel aroma attributes. Degradation of the undesired *n*-hexanal (beany off-odor) in fermented soya products has been explained by microbial activity. De novo synthesis of volatile flavor can be likewise imagined to improve the sensory quality of fermented products.

2.6 Bakery Products

With the exception of cakes and cookies, bakery products are leavened by yeasts (wheat flour) or yeast/lactic bacteria mixed cultures (rye flour). The yeasts identified were usually strains of *S. cerevisiae* (Strohmar and Diekmann, 1991; Lues et al., 1993), but also *Candida* species were reported (Torner et al., 1992; Boraam et al., 1993). The bacterial cultures were difficult to classify (for example, Strohmar and Diekmann, 1991), and even *Enterococcus faecium* was reported (Collar et al., 1992). The microbial contribution to flavor development has been difficult to establish, because the thermal generation causes a massive overlap. Volatile alcohols, acids, esters, and carbonyls were formed in the dough (Frasse et al., 1993), but the major portion was baked out in the course of the process. Only the most potent odorants, such as methional and 3-methylbutanal, can be expected to play a role in the final bread. Heterofermentative strains, such as *L. sanfrancisco*, contributed to the formation of ethanol and ethyl acetate, but did not affect the over-all flavor of the baked product (Hansen and Hansen, 1994). The sensory deficiencies of chemically leavened bread point to a more indirect contribution of the microflora. Proteases of both yeast and bacteria liberate amino acids that act as precursors for numerous thermally formed volatiles. A likewise rare and well documented example for a direct contribution of yeast metabolism was the generation of 2-acetyl-1-pyrroline, an impact component of wheat crust flavor (Grosch and Schieberle, 1991). Model experiments indicated that free L-proline was degraded by analogy to the STRECKER mechanism to 1-pyrroline which was then acetylated by oxopropanal, derived from dihydroxyacetone phosphate. Another key compound of bread crust, 2,5-dimethyl-4-hydroxy-3(2*H*)-furanone (Fig. 1.1), originated from yeast precursors (Schieberle, 1992). Boiling of an aqueous solution of low molecular weight compounds from disrupted cells yielded large amounts of the furanone. The analysis of the carbohydrates of this yeast fraction led to the conclusion, that a fructose derivative yielded the furanone upon heating. Dilution analysis of aroma extracts has revealed the contribution of further odorants to bread flavor, among them (E2)-nonenal, (E2,E4)-decadienal, acetic acid, and phenyl acetaldehyde (Schieberle and Grosch, 1994). The latter two components are most probably of microbial origin.

2.7 Nondairy Acidic Fermentations

A mixture of comminuted meat and fat, salt, sugar, spices, and curing agent can be transformed into a microbiologically stable product, fermented sausage (Salami). The typical aroma results from essential oil constituents (pepper, garlic), and again from overlapping chemical and microbial degradation reactions. In the first phase of processing, the low pO_2 eliminates contaminating pseudomonads and *Enter-*

obacteriae, and lactics and micrococci soon overwhelm the acid sensitive primary microflora. In contrast to fermented food discussed so far, minor deviations of quality and quantity of ingredients or of the processing conditions affect the sensory results (Vösgen, 1993).

In an attempt to characterize the microflora of dry fermented sausages biochemically more than 250 strains of *Lactobacillus* were isolated, among them *L. sake*, *L. curvatus*, *L. bavaricus*, and *L. plantarum* (Garriga et al., 1993). Some of the strains contained plasmids encoding for bacteriocin production. The headspace of pork and beef loin tissue inoculated with *Lactobacilli* was claimed to contain acetone, SO_2, benzene, toluene, and chloro alkanes (Jackson et al., 1992). The profiles of volatiles in the sterile packaged control samples were similar. Other authors have reported on the well-balanced aroma of sausages fermented with *Pediococcus pentosaceus* and *Staphylococcus carnosus* (Marchesini et al., 1992; Junker et al., 1993). The lipolytic and proteolytic activities of *Micrococcus* species and the triacylglycerol and protein fractions of meat seem to be well matched. High extracellular activities rendered some strains particularly suited as potential starter cultures (Selgas et al., 1993). Lipase activities of animal origin were thought to contribute to the generation of odorous short chain fatty acids (Garcia et al., 1992). Usually, a microbial contribution to aroma formation is measured by following amino N, or by evaluating 'odor' or 'flavor scores' (Perez et al., 1992), which impedes clear-cut judgements of the properties of the strains in terms of volatile flavor formation.

Moulds growing on the surface of fermenting meat (and cheese) usually belong to the genus *Penicillium*. Inoculation was carried out traditionally by the 'houseflora' of the processing plant, but, because most *Penicillia* are potentially toxigenic, selected starter strains, preferably of *P. nalgiovense*, are recommended nowadays (Geisen, 1993). More than 150 primary isolates of moulds were obtained from fermented sausages and investigated with respect to technological, sensory, and toxicological performance (Hwang et al., 1993). Two strains of *P. nalgiovense* and seven strains of *P. chrysogenum* passed the technological tests and the bioassays. No analytical data on the volatile constituents that conferred the different odor attributes were included. The application of dynamic headspace enrichment techniques has estabished the volatile fraction of salami as being the result of carbohydrate and triglycerol degradation and fatty acid catabolism together with the volatiles of the spices added (Berger et al., 1990; Croizet et al., 1992). Using various starter strains and omitting spices the contribution of the microorganisms has been recently examined in detail (Berdagué et al., 1993). Most of the volatiles statistically related to the microorganisms can be explained by metabolic pathways that have been observed in other fermented foods as well (Table 2.4). Chemical oxidation of sensitive substrates was suppressed by the low pO_2 of core and inner sections; only pentane, 1-octen-3-one, and hexanal appeared as products of fatty acid peroxidation higher concentrations. Oxidation processes will be favored in the peripheral zones of the product. The empirical use of moulds may have a chemical background in decreasing oxygen also in the outer zone to reduce rancid odor notes during prolonged maturation. The inhomogeneity of products, lots, and

Table 2.4. Impact components (selected according to known odor thresholds) of dry fermented sausage (Berdagué et al., 1993)

Compound	Quantity[x]
2,3-Butandione	90
3-Methylbutanal	153
2-Pentanone	209
2-Hexanone	28
n-Hexanal	309
2-Heptanone	57
1-Octen-3-ol	78

[x] Median, expressed in ng of n-C_{12}-an

processes leaves much room for future biotechnological input (Chasco et al., 1993).

Volatiles of vinegar were identified as higher and methyl branched homologues of acetic acid, several short chain alkanols, their corresponding esters and some other trivial carbonyls. Improvements to the process have been suggested (Fukaya et al., 1992). Modern instrumentation was applied to reanalyze the volatile composition (Blanch et al., 1992). The prokaryotic microbiology obviously restricts the metabolic diversity considerably.

2.8 Fermented Plant Materials

Cabbage, cucumber, olives, celery, peppers, and other vegetables can be preserved by spontaneous or starter-culture-induced acidification. The microbial ecology of the fermenting materials is confined to lactic-acid-forming organisms (Hammes, 1991). *Lactobacilli* adapted to meat fermentation were successfully transferred to fermenting sauerkraut, and vice versa, a sauerkraut strain outnumbered a meat-borne flora and govered the meat fermentation (Vogel et al., 1993). This interchangability of the principle substrates is accompanied by very similar odor profiles if the plant contribution is substracted. During cucumber fermentation, ethanol, propanol, glycerol, and acetic, propanoic, and butanoic acid were analyzed simultaneously by high pressure liquid chromatography (McFeeters, 1993). Less is known of the volatiles of other fermented vegetables. In contrast to other plant materials, *Pichia* and *Hansenula* yeasts dominated in aerated fermentations of olives (Gonzalez et al., 1992; Borcakh et al., 1993), and sensory results pointed towards a preference for naturally fermented olives, when *L. plantarum* was used as a pure starter (Montano et al., 1993).

In-depth analytical studies are also considered mandatory in the field of cocoa fermentation. Acetic and lactic acid bacteria dominate this process (Samah et al., 1993). Higher levels of acetic and lactic acid in treated beans lowered the chocolate flavor. The opposite results led Holm et al. (1993) to conclude that a reduction of these acids may not be sufficient to produce a desirable flavor balance, as other acids, such as citric or oxalic, also play an important role in flavor development. A series of cocoa fermentations with different cultivars, methods of fermentation and duration of post-harvest pod storage did not produce significant differences of quality as measured by the cut-test and acid concentration (Tomlins et al., 1993). Future studies should no longer ignore that some evidence for the chemical nature of chocolate flavor precursors has been accumulated. Successive enzymatic digestion of a seed protein to hydrophilic peptides by a cocoa endoprotease and a carboxypeptidase formed active precursors, as shown by heating in the presence of reducing sugars. A synthetic mixture of free amino acids according to the over-all amino acid composition of cocoa did not work (Voigt et al., 1993). This would limit the microbial effects to a protease mediated contribution, or to the creation of a chemical environment favoring plant enzyme activities.

2.9 Ethnic Food

Traditional fermented foods are especially popular in some Far East (Oriental Food) and many developing countries. A knowledgeable review listed about 70 different fermented products without aiming at completeness (Beuchat, 1987). Only a small selection of products will be discussed here with the focus on some specific flavor aspects.

2.9.1 Koji and Derivatives

The term koji refers to the starter or first phase of fermentation leading to soy sauce (Shoyu) and may be regarded as the equivalent of barley malt in western alcoholic fermentations of cereals. In Japan, equal amounts of soybean and wheat flour are fermented, soybean only in China and Korea, and fish in Indochina. About 1.2 million metric tons of soy sauce and 560 000 tons of soybean paste (Miso) are produced annually in Japan. The flavor of soy sauce, as reviewed recently (Sasaki and Mori, 1991), contains short and medium chain alcohols, acids, esters and other carbonyls common to fermented food. In addition, a large variety of sulfur compounds, alkanolides and other O, N- and S-heterocycles can be mentioned. Much of the recent work on flavor formation has been published in Japanese. Saigusa et al. (1993a, 1993b) and Kida et al. (1991) have studied rice as a starting material. The cultivation time of Aspergillus kawachii had a significant effect on the formation of higher alcohols and esters by yeast. The conventional fermentation, using steamed rice, was found inferior. For maintaining a high content of methylbutyl acetate the selection of an appropriate biochemical environment in the successive processing steps was required. The ester was not hydrolyzed in the second phase of fermentation, and it was supposed that the esterase of the starter koji was not active in the shoyu mash anymore. The *Aspergillus* species could be substituted by an addition of citric acid and a mixture of starch degrading enzymes; however, less methylbutyl acetate and higher fatty acid ethyl esters were formed in these experiments. Salted fish pose new problems due to the contents of highly unsaturated fatty acids and the lack of endogenous antioxidants. In spite of this unfavorable starting situation it was found that peroxide formation went through a transient maximum and decrease during maturation (Yankah et al., 1993). It appeared that microbial growth reduced pO_2 to a very low level to prevent oxidative spoilage. This view is supported by minor chemical and sensory differences found in aerobically and anaerobically fermented fish (Sanceda et al., 1992). However, the anaerobically produced sauce was less rancid and thus preferred to the traditional product. The long tradition of shoyu making has not yet resulted in a general agreement on processing steps and microorganisms to be used, nor on the volatile products to aim at. While one group inoculated the prefermented koji with *Pediococcus halophilus, Zygosaccharomyces rouxii* and *Candida versatilis* and emphasized 4-ethyl guaiacol and 2-phenyl ethanol (Muramatsu et al., 1992), others used *Streptococcus faecium* and *Torulaspora delbrueckii* and measured fatty acid, ethanol and methylpropanol contents (Hwang and Chou, 1991)

A protease mixture or an enzyme extract from *Bacillus subtilis* performed similarly well for flavoring miso, a cereal-based seasoning (Yoneyama et al., 1992). A recent analytical survey of 48 kinds of miso was based on adsorptive extraction and reported 4-*HO*-2(or 5)-ethyl-5(or 2)-methyl-3(2*H*)-furanone as an impact component (Sugawara et al., 1992). A total of fourteen volatile compounds explained 76% of the total statistical variation of the sensory data. The various types of natto, a solid meat substitute, are also derived from koji. Coupled gas chromatography-mass spectrometry identified more than 60 volatiles. Sulfur compounds and *O*-heterocycles can again be addressed as key contributors (Tanaka and Shoji, 1993). This analytical progress should provide a better orientation for further process developments. Miso has also been prepared from western starchy legumes, such as peas and beans, *Rhizopus*, and koji (Reiss, 1993a, 1993b). A sweet, nutty odor without 'beany' off-notes, but no analytical data on the volatile composition were reported.

2.9.2 Sake

Disregarding genuine grape volatiles, similar precursor biochemistry, microbiology, and processing steps are responsible for the wine-like spectrum of volatiles of sake (Sakamoto et al., 1993). Methylbutanol and its acetate ester were followed during isothermal fermentations (Matsuura et al., 1992). Kinetic modeling and dynamic programing calculations suggested a temperature profile of the fermention to maintain high levels of these perceived key volatiles. A benefical effect of added adsorbents of flavor formation was attributed to their hydrophobicity resulting in a decrease of pCO_2 from the supersaturated condition to the theoretical saturated concentration (Takezaki et al., 1993). The concurrent adsorption of volatile flavors in the presence of an excess of ethanol appearently did not interfere with the sensory over-all quality of the product. Sprouting rice powder was the malt replacer in a process for an alcoholic beverage designated as 'sprouting rice wine' (Teramoto at al., 1993). This fermented beverage could be the antique predecessor of modern sake. The aroma of the product was very similar to Japanese sake. In a comparative experiment, the saccharifying activity of malt was higher than that of sprouting rice. Fungi other than yeast may have grown in the mash of the early rice wines and could have modified the flavor of sake the way it was described in an recent patent (Kuwabara et al., 1993).

2.9.3 Miscellaneous Fermented Food

The microflora of fermented soy milk was similar to those of cow milk, as were the volatiles reported (Li, 1986; Murti et al., 1993b). *Lactobacillus acidophilus* and *L. bulgaricus* produced acetaldehyde, the volatile top note, and butanol, 2-butanone and diacetyl. *Streptococcus thermophilus* and *S. cremoris*, when used as pure starters, yielded a yoghurt-like product with satisfactory aroma. Hexanal was not completely removed in all products, but did not spoil the sensory over-all impression.

The aroma components of many indigenous fermented foods, such as maize (Halm et al., 1993) and sorghum dough (Hamad et al., 1992), or cassava (Meraz et al., 1992), are much less characterized. In general, the microflora and, as a result, the volatile flavor compounds identified, overlap considerably with the thoroughly investigated western world counterparts. Biochemical data are often confined to acidification rates or changes of concentrations of nutritionally significant aroma precursors, such as essential amino acids (Hamad et al., 1992). As mentioned above (Vogel et al., 1993), there appears to exist a remarkable degree of freedom in exchanging fermentation substrates and fermenting microorganisms. The examples of a soy milk fermentation using cultures from fermenting maize (Nsofor et al., 1992), and a rice flour fermentation using lactic acid bacteria from fermenting fish (Lee et al., 1992b) demonstrated that acceptable or pleasant novel odor attributes can be created under suitable conditions. In order to keep pace with the modern trend of developing continuous processes, attempts toward an immobilization of strains of oriental fermented food were made (Tamada et al., 1992; Iwasaki et al., 1992). Lactic acid generation was frequently the only chemical parameter reported. Flavor compounds emanating from a traditional Japanese apricot liqueur fermented with alginate entrapped yeast were described (Takatsuji et al., 1992). Immobilized glutaminase and yeast were applied in a two-step process for the production of soy sauce (Kanematsu et al., 1992). Taste was improved by the enzymatic conversion of glutamine to glutamic acid, and odor gained from the yeast associated generation of 3-methylbutanol, 2-phenyl ethanol, and 4-vinyl guaiacol. These achievements must not obscure our judgement of the fact that major insight into biochemistry and microbiology of flavor formation in empirically fermented food is lacking. As long as species, such as *Proteus mirabilis, Pseudomonas aeruginosa* or *Enterobacter cloacae* are cited, and aroma improvements are stated in a non-objective manner ('more odorous', etc.), the procedures must remain doubtful. Research needs are now well recognized and have, for example for traditional African foods, been recently formulated (Sanni, 1993).

2.10 Aroma Enhancement by Mixed Cultures

The biodegradation of complex food constituents in traditional processes often benefits from the presence of a microbial community. Continuous-flow enrichment techniques showed that various types of interactions were encountered. Without going into the classification of mechanisms and effects it is evident that a single strain cannot be expected to introduce all of the desirable biochemical and operational features into an optimal process. For example, heat treatment, soaking or extrusion of the raw materials may facilitate the access of the biocatalyst to the substrate and remove inhibitors, but a suitable microflora may afford the same with less expense.

Work compiled in Table 2.5 showed that the production of ethanol and of other volatiles from demanding substrates could be greatly stimulated by mixing biocatalysts. How close the biochemical relationships can be, has been experienced in the case of rice straw fermentation: a coupled system performed better than two stages with separate hydrolysis and fermentation (Deshpande, 1992). Raw starch was converted into ethanol by a complex, co-immobilized community of prokaryotics

Table 2.5. Process improvements by mixed biocatalysts in ethanol dominated fermentations

Biocatalysts	Advantage	Reference
A. niger/S. cerevisiae	Ethanol yield	Ohta et al., 1993
Cellulase/*S. cerevisiae*	./.	Deshpande, 1992
Pichia stipidis/S. diastaticus	./.	Laplace et al., 1993
A. awamori/Rhizopus japonicus/ Zymomonas m.	./.	Lee et al., 1993
Hansenula or Kloeckera/*S. cerevisiae*	Volatiles production	Zironi et al., 1993
L. plantarum/S. cerevisiae	Sherry aroma	Suarez and Agudelo, 1993
Mixed yeast culture	Beer aroma	Demuyakor and Ohta, 1993
Enterobacter/Yeast	Lambic beer	Martens et al., 1992

and eukaryotics (Lee et al., 1993). Interactions of the metabolic routes to volatile flavors are less clear, but have been demonstrated by comparing pure, mixed, and sequential cultures of *Saccharomyces* and wild yeasts occurring in grape musts (Zironi et al., 1993) The fermentation of Belgian lambic beer was spontaneously initiated by enterobacteria and yeasts (Martens et al., 1992). Model experiments with bacterial isolates showed that most of the ethanol was produced by the yeasts. Bacterial species that died off during maturation of the beer produced most of the other volatiles. In contrast to these findings, the addition of lactic acid bacteria, isolated from a commercial malt whisky distillery, to a fermenting mash was not favorable under controlled conditions (Makanjuola et al., 1992): Reduced yields of ethanol, reduced carbohydrate turnover and yeast growth, increased acidity and foam production were important drawbacks of the bacterial contaminants. On the other hand many whiskies surely owe part of their specific, esteemed flavor to in-house strains of lactics.

Some mixed starters of the dairy field were mentioned above. Heterogenous, sometimes immobilized inocula were used for the fermentation of cream or cheese whey permeate (Prevost and Divies, 1992). Kefir grains represent a natural form of a co-immobilizate (Leroi and Pidoux, 1993). CO_2 and some low molecular weight acids, excreted by the yeast, stimulated the growth of *Lactobacillus hilgardii* isolated from the grains. N-compounds were not involved in this effect. A novel approach to study low molecular weight compound mediated interactions was a fermenter with two compartments separated by a dialysis membrane (Klaver et al., 1992). One compartment was inoculated with the citrate utilizing, diacetyl producing *Leuconostoc* or *Lactococcus* strain. As the *Lactobacilli* in a second 'product' compartment did not convert citrate and lacked diacetyl reductase, a stable system for the production of diacetyl flavored buttermilk was established. The same dialysis reactor, inoculated with the acetaldehyde forming *Lactobacillus delbrueckii*, was used to produce a mildly flavored yoghurt. The concept is versatile and may be extended to many other fermented foods. More detailed chemical and biochemical evaluations will be needed. This also holds true for mixed milk-flour or sour dough fermentations (Lazos et al., 1993; Javanainen and Linko, 1993). Microbial performances in complex media may have been accelerated by the simultaneous utilization of physiologically related substrates for a long time (Babel et al., 1993). Formation of primary metabolites is linked to cell growth and, thus, limited either by energy or reducing equivalents. Removal of one of these bottlenecks by combining appropriate substrates can therefore be expected to increase yields. Practical experience indicates that not only mixed cultures, but also mixed substrates can activate existing metabolisms including degradation reactions.

2.11 Conclusions

Modern food biotechnology has emerged from empirical, self-regulated fermentations and is still centered on preventing nutritious food from microbial spoilage by the production of acid, ethanol and further inhibitors. Other aspects of food quality, such as flavor, color, texture or digestibility are observed to improve in these integrated processes. Though certain products, after fermentative refinement, are directly used as seasonings, it is usually difficult to develop a concerted biotechnology of an aroma compound from the complex biochemistry of an empirical process. Exceptions to this rule are those volatiles of which at least one facultative metabolism was known, for example, acetaldehyde, diacetyl, certain fatty acids and fatty acid esters.

Gathered experience should now flow into novel bioprocesses. Keywords are chemical environments including precursors and gases, growth factors and interactions, biosorbents, immobilization, metabolic screening, and bioreactor design. The dialysis reactor, for example, has already been successfully applied in enzyme technology. Moreover, the aroma-forming potential of the classical food strains and related wild strains is far from being fully exploited.

3 Why Novel Biotechnology of Aromas?

As outlined in the preceding chapter, the products of food biotechnology continue to command a significant portion of the food market. More recently, bioprocesses have been developed for nonvolatile flavor compounds to be used as food additives. Manufactured on an industrial scale are various organic and amino acids, 5'-nucleotides, and several types of hexose syrups (Tombs, 1990). On this broad background increasing interest in bioprocesses for the generation of volatile flavors in both academia and industry can be noticed. The multiple and different reasons of this interest are mainly linked with:

- problems of supply from agriculture,
- the present legal restrictions,
- the high stereochemical requirements, and
- the dramatic advances in cell biochemistry and bioengineering.

3.1 Aromas from Living Cells

In contrast to repeated predictions, the trend to natural flavors is unbroken. Food flavors share about 2 billion US$ in the worldwide 5 billion US$ market of volatile flavors and fragrances. Natural aromas account for 60% and 40% respectively, of the total sales in the European Union and in the United States; the respective figures for Japan and the remaining countries are 10% and 5%. Sources for natural aromas are:

- extracts (oleoresins, ethanolic percolates, concretes, absolues),
- distillates (of extracts or from plants; essential oils),
- pressed oils (mainly *Citrus*),
- concentrates (mainly of fruit juices),

and fractions or chemically pure compounds obtained by repeating the above techniques, or by chromatography etc (Matheis, 1989). Volatile emissions of the processing of fruits are recovered by combined heat exchanger or rectification/condensation units, and the aroma concentrates can be used for reconstitution prior to packaging, or to flavor other food products (Kollmannsberger and Berger, 1994). Aromas generated during baking or other thermal processes must be removed from the waste air from environmental reasons, but these fractions are not suitable for reincorparation: The end product has never contained these volatiles, and the sensory balance would be lost. However, 'today's trash might well be tomorrow's treasure' (Reineccius, 1992) as shown by the example of dimethyl sulfide. Once distilled off from mint oil and discarded, it is now worth 800 US$ per pound as a natural flavoring. In addition to physical processes, natural aromas may also be obtained by fermentation. The heart of each biotechnology is the biocatalyst. In the flavor field

- crude and purified enzymes, mixtures of enzymes,
- intact prokaryotic or eukaryotic cells, or
- plant cells as callus or suspension cultured cells

have been used. On the whole, biotechnology adds just one more source of flavors to the existing ones. Agricultural sources, however, may suffer from problems of soil, climate, pests, and politics. These inherent instabilities are major drawbacks in a growing market. Character impact components are often present in plant sources in minute amounts and conjugated forms, thereby creating another problem: economic isolation. Hence, independent and constant supply are prime motives for adopting new biotechnology.

3.2 The Legal Situation

The definitions used by the US Food and Drug Administration and the mandatory guideline of the previous Council of the European Communities classify fermentation aromas, with few restrictions, as being natural aromas. As the average consumer links naturalness with safety and purity, these regulations are easily interpreted as strong marketing arguments.

In the United States, flavors produced by fermentation (and/or processing) are natural if the starting materials were natural. Irrespective of its origin, a substance or mixture and the intended use is assigned a GRAS status after passing a review by an expert panel of the FEMA (cf. Sect. 1.3). Grey zones exist for the use of 'natural' catalysts, for example acids or bases derived from biotechnology. This 'soft chemistry' intentionally alters chemical structures and has never been accepted in Europe. The FDA has indicated that it will require clearance by either a food additive regulation or by a regulation confirming GRAS status for products with conceivable risks. As aroma compounds are volatile by definition, efficient downstreaming processes can be applied to assure the exclusion of, for example, recombinant nucleic acids or other cell debris (cf. Sect. 10.3). In these terms, any concern over the biocatalytical system, particularly if immobilized, could be cleared up. Of course, the application of food strains (cf. chapter 2) will facilitate approval by the authorities. More details may be taken from a recent review on biotechnologically produced food ingredients (Appler and Giamporcaro, 1991).

The European guidelines (88/388/EWG of June 22, 1988; 91/71/EWG and 91/72/EWG of January 16, 1991) define natural aroma compounds as isolated by 'physical, enzymatic, or microbiological processes or traditional food preparation processes solely or almost solely from the foodstuff or the flavoring source concerned'. Though adopted from previous, national flavor regulation, these formulations leave space for interpretation. For instance, autocatalysis of the reaction of a natural acid by its equilibrium protons would be acceptable, but not an esterification that would not proceed at an economic rate without added citric acid. The isolation of a flavor aldehyde from a fermentor broth using hydrogen sulfite would not be permissible, but the removal of the same compound as an off-flavor creating impurity should not pose a problem, nor does the use of genetically engineered biocatalysts or proven pathogens. The draft proposal for the coming council regulation on novel foods did not mention aromas (92/C190/4). No obligatory regulations of genetically modified food exist in any European country.

In view of this legal uncertainty the flavor industry has set its own high standards (IOFI, 1991). The guidelines of the International Organization of the Flavor Industry demand accordance of the bioprocess with Good Manufacturing Practice and with the general principles of hygiene of the Codex Alimentarius. All bioflavors must comply with national legislation, and safety must be 'adequately established'. This product-oriented industrial point of view reminds one of the GRAS procedure and should provide sufficient protection of the consumer's health.

3.3 The Potential of Biocatalysts

If the only justification of a bioprocess is that its product(s) may be labeled as natural, it is unlikely that the process will be economically viable in the long term. However, biocatalysts can accomplish reactions that simply cannot be carried out by chemosynthesis. If a chemosynthetic route is chosen and results in a mixture of products or isomers, a subsequent separation may be more costly than an established bioprocess. Not only can trace impurities of chemosynthesis falsify the sensory character of a product, but sensory or generally physiological activity is often dependent on an exact stereochemical structure (cf. Sect. 1.5). Enantiomeric forms of a chiral character impact component may exhibit different odors. This holds for mono-, sesqui-, and norterpenoids (Fig. 3.1), and also for lactones, other heterocycles, methyl branched aliphatics, and many other volatiles. Both forms of an enantiomeric pair can occur in essential oils and coin the respective odor, for example the enantiomer α-pinenes in caraway ($1R$) and parsley seeds ($1S$), or the limonenes in dill herb ($4R$) and hop oil ($4S$ dominating). The well-known influence of chirality on odor perception was recently investigated for camphoraceous, urinous, ambergris, and grapefruit odorants (Chastrette et al., 1992). Molecular structures of superimposed conformations of reference and sample constituents were compared by molecular modeling, and some correct predictions, but also limitations were presented. As odor recognition is chiral, the most adequate syn-

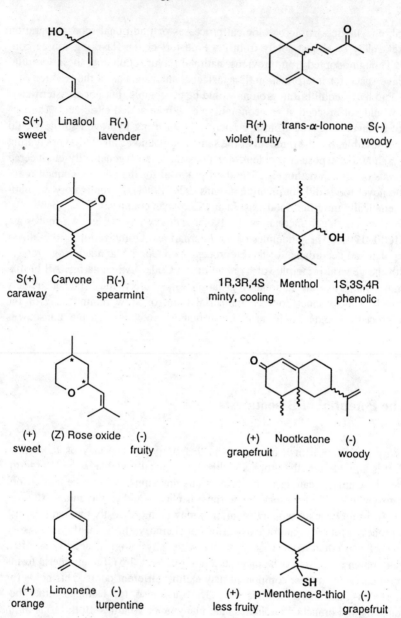

Fig. 3.1. Effect of chirality on odor character of volatiles

thetic approach would be using chiral (bio)catalysts. Though biocatalysts are not free from inherent drawbacks, some advantages, such as ecological compatibility or multistep synthesis cannot be matched by a (chiral) chemical catalyst (Table 3.1). Biocatalysts should, therefore, be thought of in terms of providing a complementary set of options. The reduction of geraniol (Fig. 3.2), and the oxida-

Table 3.1. Properties of biocatalysts (L-polyamides) compared to nonbiochemical catalysts

Advantages
High specificities (chemo, regio, stereo-)
High reaction velocities at low molar fractions
Ecologically compatible agents and reaction conditions (medium, energy)
Simple process, process control and work-up (enzymes)
Cofactor/cosubstrate regeneration (whole cell)
Multistep syntheses possible

Disadvantages
Low stability of catalytic activity
Problems of cost (pure enzymes) or availability
Bioreactor and tedious work-up required (whole cells)
Highest catalytic activity in water
(Enantio)Selectivity and side reactions difficult to predict

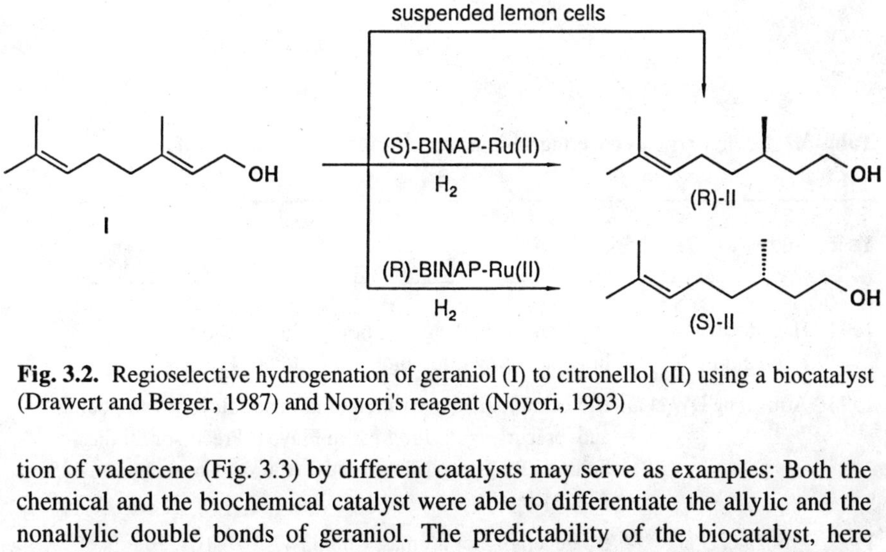

Fig. 3.2. Regioselective hydrogenation of geraniol (I) to citronellol (II) using a biocatalyst (Drawert and Berger, 1987) and Noyori's reagent (Noyori, 1993)

tion of valencene (Fig. 3.3) by different catalysts may serve as examples: Both the chemical and the biochemical catalyst were able to differentiate the allylic and the nonallylic double bonds of geraniol. The predictability of the biocatalyst, here cells of *Citrus limon*, leaves a lot to be desired. The chemical catalyst, on the other hand, is highly sophisticated and the reaction requires free hydrogen gas at high pressure. The biocatalyst similarly converted isomer farnesols to (E6)-2,3-dihydrofarnesol, while bis-homo-geraniol was inert in the chemical system. The industrial importance of nootkatone is underscored by many efforts toward its (bio)synthesis; an undefined plant oxidase is said to have finally solved the problem of selective production on an industrial scale. More examples showing in-

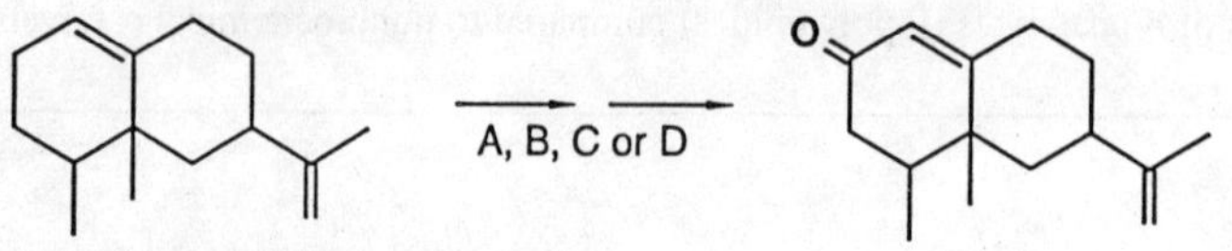

Fig. 3.3. Regioselective oxidation of valencene to yield nootkatone, one of the character impact components of grapefruit
A: *tert*-Butyl chromate, *tert*-Butyl peracetate, or photooxidation in low yields,
B: 'Soil bacterium'; 12% yield + 7,11-epoxide, 11-ketone, and other products (Dhavlikar and Albroscheit, 1973)
C: suspended cells of *Citrus paradisi*, 66% yield in 6 hours, the immediate precursor 2-hydroxyvalencene being the only detectable volatile side product, D: crude isolate from industrial *Citrus* waste (proprietary data)

creasing recognition of the potential of biocatalysts in the flavor field can be found in recent reviews (Table 3.2). Many authors have discussed one class or various groups of aroma compounds under quite different aspects. For an introduction to the older literature and a description of some processes already operating the reader is referred to these reviews. Novel analytical methods and biotechnical advances have been reported since then which will broaden and facilitate the use of biocatalysts for the production of volatile chemicals.

Table 3.2. Review articles covering various aspects of the biogeneration of aroma

Year	Author(s)	Source
1994	Hagedorn S Kaphammer B	Biocatalysis in the generation of flavors and fragrances In Annual Rev Microbiol, 48, 773-800, 156 refs.
1993	Armstrong DW et al	Biotechnological derivation of aromatic flavour compounds and precursors In: Progress in Flavour Precursor Studies, Schreier P, Winterhalter P (eds) Allured, Carol Stream, 425–438, 21 refs.
1993	Fronza G et al	Biogeneration of aromas, Chimia 47, 107–109, 16 refs.
1993	Linton CJ Wright SJL	Volatile organic compounds: Microbiological aspects and some technological implications, J Appl Bacteriol 75, 1–12, 152 refs.
1993	Winterhalter P Schreier P	Biotechnology, Challenge for the Flavor Industry In: Flavor Science, Acree TE, Teranishi R (eds) ACS Wash DC, 225–258, 240 refs.

| 1992 | Belin JM et al | Microbial biosynthesis for the production of food flavours, Trends Food Sci Technol 3, 11–14, 48 refs. |

1992 Belin JM et al   Microbial biosynthesis for the production of food flavours, Trends Food Sci Technol 3, 11–14, 48 refs.

1992 Berger RG et al   Naturally-occurring flavours from fungi, yeasts, and bacteria In: Bioformation of Flavours, Patterson RLS Charlwood BV MacLeod G Williams AA (eds) Royal Soc Chem Cambridge, 21–32, 52 refs.

1992 Bigelis R   Flavor metabolites and enzymes from filamentous fungi, Food Technol 46, 151–161, 39 refs.

1992 Janssens L et al   Production of Flavours by Microorganisms, Proc Biochem 27, 195–215, 248 refs.

1992 Farbood MI   Micro-organisms as a novel source of flavour compounds Biochem Soc Trans 19, 690–694, 18 refs.

1992 Quehl A
       Ruttloff H   Mikrobielle Produktion von Aromastoffen unter besonderer Berücksichtigung von Fruchtnoten, Nahrung 36, 159–169, 63 refs

1992 Romero DA   Bacteria as potential sources of flavor metabolites, Food Technol 46, 122–126, 35 refs.

1991 Berger RG   Genetic Engineering Part III: Food Flavors In: Encyclopedia of Food Science and Technology, Hui YH (ed) Wiley New York 1313–1320, 59 refs.

1991 Seitz EW   Flavor Building Blocks In: Biotechnology and Food Ingredients, Goldberg I Williams R (eds), van Nostrand Reinhold New York, 1991, 375–391, 33 refs.

1991 Suzuki M   Fragrance and off-flavors produced by microorganism, Koryo Koizumi T 171, 61–81, 108 refs.

1989 Bell ERJ
       White EB   The potential of biotechnology for the production of flavours and colours for the food industry, Int Ind Biotechnol 9, 20–26, 98 refs.

1989 Hanssen H-P   Fermentative Gewinnung von Duft- und Aromastoffen aus Pilzkulturen, GIT Fachz Lab 33, 996–1003, 9 refs.

1989 Kinderlerer JL   Volatile metabolites of filamentous fungi and their role in food flavour, J Appl Bacteriol Symp Suppl 133–144, 60 refs.

1989 Schreier P   Aspects of biotechnological production of food flavors, Food Revs Int 5, 289–315, 111 refs.

1989 Grüb H
       Gatfield JL   Generation of flavor components by microbial fermentation and enzyme technology In: Flavor Chemistry of Lipid Foods Min DB Smouse TU (Eds) AOCS, Champaign, 367–375, 9 refs.

4 Laboratory Requirements and Techniques

This chapter will provide the beginner with some practical introduction and experience gathered with the laboratory-scale cultivation of aroma producing microorganisms and plant cells. Basic and specific laboratory equipment and the infrastructure required will be described. For a detailed discussion of general biochemical, microbiological, or biotechnological principles a great number of textbooks are available; advice direct from an experienced microbiologist/biotechnologist will help to get started more rapidly.

4.1 Laboratory Equipment

Aroma generation by a living cell depends, like any other physiological process, on exogenous parameters, such as temperature, water activity, partial pressure of gases and a full array of nutrient chemicals. Bacteria and, to a lesser extent fungi, can adapt to extreme conditions that direct metabolites to volatile products (Table 4.1). In practice, a selected strain has to be subcultured several times and adjusted stepwise to production conditions. Biological status (activity, age, spore forms) and amount and preparation of the inoculated cells (Abraham et al., 1994) determine the lag-phase of a static culture prior to the phase of constant maximal growth (exponential, linear phase). After a certain cell density has been reached, lack of oxygen or other substrates or the accumulation of toxic metabolites forces the cells into a 'stationary' phase of growth that precedes cell death. Volatile flavors are not necessarily typical idiolites (cf. Sect. 5.6); they can also be formed in the phase of active growth (Abraham et al., 1993).

Clean and sterile manipulations serve both the microbial operation and the operator. Irrespective of the pathogenic potential (or GRAS status) of the strain, the use of a laminar flow bench is recommended to exclude nonsterile air, the major source of contaminants. Other pieces of apparatus, such as a switchable burner, inoculation tools, sterile pipettes and glass, plugs, and desinfectant should be stored closeby, but not inside the laminar flow bench. Nutrients are sterilized in storage vessels with a plugged, lateral side arm through which petri plates can be filled directly. Technical and analytical balance, pH-meter, autoclav, stirrer, and packaging materials are required. Bioreactors filled with nutrient medium and fully equipped with storage solution vessels can be sterilized in situ without too much

Table 4.1. Basic physiological data [a] of aroma generating cells

	Bacterial	Yeast	Plant
Diameter [μm]	~1	< 10	20–200
Surface/volume ratio [m^{-1}]	6×10^6	6×10^5	6×10^4
Respiration rate [μgO$_2$ ×mg^{-1} h^{-1}]	10^3	10^2	0,5–4
Generation time [min]	10	10^2	10^3
Maximum cultivation temperature [°C]	60	45	30
Arginase induction [ΔA]	10^2	10	-[b]

[a] Typically determined under standard growth conditions
[b] no data available

inconvenience up to a working volume of 4 to 5 liters; inlet and outlet openings have to be protected by sterile filters (membrane or homemade glass wool filter) to allow for pressure compensation; parts with different coefficients of expansion should be assembled loosely. Enough time must be provided for cooling down to prevent sensitive electrodes from being damaged. Thermostatted incubators and gyrotory shakers receive the small scale incubation vessels.

Continuous microscopic monitoring of the growing cells is often neglected. Standard protocols combined with staining, or phase contrast microscopy help one to detect contaminants. Particularly informative are vital stains using fluorescence microscopy to determine the viability of the cells at a glance. Surface colonies can be monitored using a stereo microscope. Additional equipment depends on the type of research to be carried out and may comprise photometer, other spectroscopy, centrifuges, and chromatography (see below).

4.2 The Biological Materials

Strains may be derived from one's own isolates, from mother tissues (plant cells) or ordered from commercial collections, such as The American Type Culture Collection (bacteria and fungi) or the Centralbureau voor Schimmelcultures (yeasts and other fungi). Though there ist no general optimal method to maintain the strains properties, low temperature and darkness are considered important. Bacteria may be preserved below – 30 °C or lyophyllized. Yeasts and fungi lose viability to some extent from all techniques and should, at certain intervals, be subcultured on full media before entering the next storage period. Fungal spores may be kept on agar tubes and refrigerated. Even with great care, degeneration (of overproducers) cannot be completely excluded (Tiefel and Berger, 1993). The preservation of plant cells is still a subject of controversy. Mineral or vegetable oils have been applied to reduce the access of oxygen to surface cultures. As the growth of cells of higher plants as surface (callus) cultures is comparatively slow anyway (Table 4.1), it is usually sufficient to maintain stock cultures at + 10 °C.

4.3 Laboratory-Scale Cultivation

Volatile metabolites are, with very few exceptions (lower sulfides, indol, buta-nol/acetone), produced by facultative anaerobics (cf.chapters 2 and 5), or obligate aerobics. U-shaped glass tubings, filled with a water barrier, create microaerophi-lic conditions in fermentation vessels. Two dimensional (surface) cultivations are proven on an technical scale (citric acid). Three dimensions (submerged culture), however, save space and costs, and are much easier to operate. Oxygen must be introduced into the liquid bioprocess medium through porous sinter glass plates, or porous metal or polymer tubings. Mechanical agitation increases the residence time and dissipation of the air bubbles in an attempt to reach the (low) limit of oxygen saturation (ca. 1 mmol L^{-1} H_2O, 37 °C). The gradient of oxygen concen-tration, the area of transfer, and the transfer coefficient as affected by the activity of the dissolved nutrient chemicals and the cells themselves are crucial for the oxygen supply of the cells. The pressure drop caused by the sterile filters and waste air condenser improves the solubility of oxygen and the microbial process safety, two welcome benefits. Fortunately, most of the known, aroma producing cells belong to taxa with comparably lower respiration rates rendering oxygen transfer a less critical factor.

Laboratory-scale bioreactors up to a working volume of 15 to 20 liters are offe-red with exchangable glass vessels and various stirrers, and controls for agitation, medium temperature, pH, and sometimes antifoam (chemical or mechanical). Lar-ger volumina vessels are manufactured of stainless steel to permit in situ steriliza-tion. Septa or sterile locks for adding or removing gaseous or liquid chemicals are included for feeding, sampling, or harvesting.

The question of implementing further controls and probes should be answered depending on the character of the biocatalyst. Measurement and control of the partial pressure of oxygen makes sure that supply has been sufficient at any time. Better than cell mass related parameters the oxygen consumption indicates the physiological status and activity of a cultured cell. On an industrial scale, oxygen control assists in improving yields, and in adjusting to the real needs (pumps, fil-ters, antifoam). As a state of partial anaerobiosis prevents a respiratory depletion of primary intermediates, secondary pathways leading to formation and accumula-tion of volatile flavors can be expected to become dominant. In fact, an oxygen pressure close to the critical value can stimulate flavor formation. The effect of the partial pressure of CO_2 on yeast volatiles was discussed above (cf. Sect. 2.2). Aeration is bound to strip a bioreactor broth, and CO_2 is among the volatiles re-moved. To control CO_2 losses, it might be advisable to equip the bioreactor with the respective probe and electronics. The redox potential of the broth runs usually parallel with pO_2. Reducing agents have been reported to stimulate the growth of *Bacillus, Acetobacter,* and yeast strains. Their interplay with nutrient constituents and metabolites secreted in the course of the cultivation can be followed with an rH-electrode (or better with the normalized potential, E_h). Turbidity (optical density) or light scattering are indicators of cell growth. Both techniques possess linear correlations only in the less important range of low cell densities. Cell growth within the optical path and interference from dissipated gas bubbles represent further obstacles. Recent developments (cf. chapter 10) promise to close this serious analytical gap.

The paramagnetic properties of O_2 and the strong infrared absorbance of CO_2 are the physico-chemical principles of *on line* gas sensors mounted in the waste air stream. Total volatile carbon can be measured *on line* using a portable flame ionization detector. Volatile flavors that are stripped according to their vapor partial pressures are conveniently trapped on the lab scale in cooled organic solvents or on polymer adsorbents. It is an important safety precaution to minimize the flow restriction thus generated. Accumulation time as such and sampling time as a function of length of the connecting tubing and the flow velocity of the waste air (or other media in general) means that *on line* measurement lags behind. The time dependent variables (concentration of volatiles or other chemicals) are no longer measured in real time, but as a time integral. This suggests a careful adjustment of the cycle time of sampling. Blue-light-induced changes of gene expression of plant cells, for example, were observed within fractions of a second. The measuring device, consequently, should have the same time resolution . In liquid samples, one should face the existence of continuing biochemical processes until chemical inhibition or thermal treatment have brought them to an end.

Continuous cultivation with its inherent risks of contamination should be reserved for the investigation of physiological parameters under otherwise constant, defined conditions. The ever-changing concentrations of substrates and products in batch cultivation produce sets of parameters that cannot be controlled in a stringent sense on a laboratory scale. With the flavor-producing higher fungi, continuous processing is largely facilitated by the strongly acidic production media (Hädrich-Meyer and Berger, 1994). An economical advantage of continuous culture can emerge from the permanent selection of strains preferable in terms of acid or product tolerance, or simply in terms of growth and, thus, productivity in the production medium (Kapfer et al., 1991). Variants of continuous culture, such a turbidostat, chemostat, fed-batch with or without cell recycle have been reported for aroma producing cells and will be discussed in the following chapters.

4.4 Nutrient Media

Minimal and full media can be distinguished. While the minimal media contain an organic carbon source and essential minerals and vitamins, full media are composed so as to maximize growth rates. High yields of volatile flavors are usually achieved with a high C/N-ratio plus metabolic precursors, stimuli, or inhibitors. Full media for growing bacteria, yeast, or fungi often contain ill-defined ingredients, such as yeast or malt extract, or various peptones. Lot to lot deviations and high costs have induced the development of chemically better defined media. Solidifiers and buffers or indicators may complete the recipe. If the formation of a volatile product has not been linked to growth or is dependent on overflow mechanisms, maximum yields were reported for minimal media. Fungal cells, for example, may grow well in a medium containing a single amino acid as a N-source and a triacylglycerol as a sole C-source. High concentrations of volatile catabolites of fatty acids were produced as a result (Tiefel and Berger, 1993). Photoheterotrophic plant cells are particularly complicated to grow in vitro. An overview of culture conditions affecting growth and formation of volatile flavors by plant cells is compiled in Table 4.2 (cf. chapter 9).

Table 4.2. Factors affecting plant cell culture

Biological
Origin, size, preparation of inoculating cells
Differentiation of specialized tissues
Conditioning of the medium
Species and strain

Chemical
Carbohydrate(s)

Minerals:	NH_4NO_3, KNO_3, $CaCl_2$ aq, $MgSO_4$ aq, KH_2PO_4, $FeSO_4$ aq, $MnSO_4$ aq, H_3BO_3, $ZnSO_4$ aq, KJ, Na_2MoO_4 aq, $CuSO_4$ aq, $CoCl_2$ aq
Others:	Na_2EDTA, casein hydrolysate, vitamins, phytoeffectors, gases, precursors, antibiotics, antifoam agent, pH corrigentia; leaf extract, juice, coconut water . . .

Physical
Light (λ, duration/intervals)
Temperature
Agitation (v and intensity)
Design of bioreactor, stirrer, aeration

4.5 Analysis of Nonvolatiles

Almost all of the whole cell based bioprocesses rely on the turnover of nonvolatile constituents of the nutrient medium. Non or less volatile metabolites formed are typically polyvalent alcohols, fatty acids, and organic acids. Membrane sampling devices separate these solutes from intact cells, cell debris, and polymer material. The samples are then analyzed *off-line* using enzyme kits, spectrophotometry, or liquid chromatography often using serially coupled detectors. The over-all chemistry of educts, precursors, and products will determine which of these routine techniques should be chosen for a specific application. Addition of internal or external standards combined with computerized integration of peak areas give a satisfactory quantification of chromatographic results.

An increasing selection of chiral stationary phases for liquid chromatography permits nonvolatile enantiomers to be separated directly. Multiple noncovalent interactions, such as hydrogen bonding, π-π interaction, hydrophobic interaction and steric inclusion can act together in modified cyclodextrin-bonded phases for superior selectivity. Chiral fused silica capillary columns, as they are applied in capillary electrophoresis, add further resolution power and can be coupled with computerized diode array detectors. Liquid chromatography can be interfaced with mass spectrometry to identify the separated enantiomers. As the legal regulations demand bioprocess media to be composed of natural constituents if a bioproduct is to be labeled as natural, the chiral analysis of the precursors and substrates will gain increasing importance.

4.6 Analysis of Volatile Bioflavors

Analytical evaluation of the success of a bioprocess for the production of volatile flavor depends on the effectiveness of the isolation technique. Separate or simultaneous distillation/extraction steps and headspace enrichment are the most popular options (cf. Sects. 1.1; 2.5). Supercritical fluid extraction, favorably combined with supercritical fluids as mobile phases in a subsequent chromatography, is applicable to a broad spectrum of target compounds. Once isolated, aroma concentrates should be routinely examined by one of the available GC-O methods (cf. Sect. 1.2; Abbott et al., 1993). Coupled gas chromatography-mass spectrometry and, if the product concentration permits, coupled gas chromatography – Fourier transformed infrared spectroscopy can be used to identify bioflavors (Full et al., 1991). For metabolic studies using radiolabelled substrates the specialized instrumentation of an isotope laboratory will be required. A sophisticated, powerful analytical supplementation is radio-GC with capillary columns (Reil et al., 1992).

4.6.1 Chirospecific Gas Chromatography

Authenticity control of bioflavors is the domain of enantioselective inclusion gas chromatography (Table 4.3). More than 30 chiral liquid phases with cyclodextrin chemistry have been described (Hener and Mosandl, 1993). The volatile compo-

Table 4.3. Enantioselective capillary gas chromatography for the analysis of chiral volatile flavors

Compound(s)	Reference
α- and β- ionones	Braunsdorf et al., 1991
rose oxides, linalool, citronellol, carvone	Kreis and Mosandl, 1992
4- and 5-alkanolides, α-ionone	Lehmann et al., 1993
theaspiranes	Schmidt et al., 1992
4-alkanolides	Werkhoff et al., 1993a
flavor and fragrance chemicals	Werkhoff et al., 1993b
4-alkanolides	Vorderwülbecke et al., 1992
5-alkanolides	Nago and Matsumoto, 1993
epoxy alkanols	Hamberg, 1992

und under investigation must not be subject to racemization due to acid-catalyzed or structure-induced isomerization. An efficient sample clean-up facilitates enantioseparation. The order of elution has to be unambiguously determined for each pair of isomers, and even an inversion has been observed for homologues eluting from the same column (Mosandl, 1992). A recommended instrumental configuration consists of an achiral capillary precolumn coupled to a chiral high efficiency column, followed by mass spectrometric analysis using single or multiple ion monitoring for maximum sensitivity. Results obtained with processed fruits indicate that many bioflavors will survive prolonged incubations in an acidic, aqueous environment without racemization (Lehmann et al., 1993). Bioreactions are prone to deliver a single enantiomer form of the target compound (cf. Sects. 3.3; 7.4). This pure compound may, however, exhibit the opposite stereochemistry than the one desired. Depending on the selectivity of the biocatalyst, racemic mixtures of structurally stable chiral volatiles may also be produced (Braunsdorf et al., 1991). Among the racemic compounds found in living plants were many chiral monoterpenoids, such as nerol oxide, α-pinene, and limonene. A high variation of the distribution of enantiomer theaspiranes was reported for different natural sources (Schmidt et al., 1992). In raspberry cultivars, traces of 4-decanolide found were racemic, while the (R)-(+)-configuration dominated in higher actual concentrations (Werkhoff et al., 1993). In dairy products, as an example of fermented foods, racemic 4-alkanolides (Vorderwülbecke et al., 1992) and R-(+)-5-alkanolides (Nago and Matsumoto, 1993) may indicate a parallel formation of volatiles on chemical and biochemical routes. Structural elucidation of compounds containing more than one asymmetric carbon may require a derivatization protocol that maintains the genuine configuration or inverts the configuration in a predictable, quantitative mode (Hamberg, 1992). Diastereomer derivatives can be analyzed on achiral columns. Further progress in the classification of bioflavors can be expected in the field; the above discussed limitations need to be taken into account.

4.6.2 Stable Isotope Analysis: IRMS and SNIF-NMR

Analysis of volatile flavors that are devoid of chiral carbons is based on the natural abundance of stable isotopes (H, C, N, O, S). Current investigations take advantage of the fractionation of the isotopomere forms of CO_2 during the carboxylation of phosphoenolpyruvate (C_4 plants, discrimination of ^{13}C -16 to -10 ‰) and carboxylation of ribulose-1,5-bisphosphate (C_3 plants, discrimination of ^{13}C -32 to -24 ‰). Terrestrial plants possessing the Crassulacean acid metabolism couple C_3 and C_4 assimilation and metabolite transfer within single cells (CAM plants, discrimination of ^{13}C -30 to -10 ‰). Physical diffusion and dissolution processes, growth close to industrial combustion of fossil carbon sources, and the chemical fractionation of the CO_2/HCO_3^- equilibria overlap with the enzymatic discrimination (PEP carboxylase: -2 ‰). Isotope effects smaller than 1 ‰ should be skeptically evaluated, though the reproducibility of a modern mass spectrometer may reach ± 0.02 ‰, and with automated combustion ± 0.01 ‰ or better. Not all the leaves of an individual plant, not even all the parts of a particular leaf show the same $\delta - ^{13}C$ values. Some rules can help us to interpret the results:

- The more bonding to an atom, the greater the trend of the heavier isotope to accumulate at that site.
- Physical and chemical processes transform the heavier isotope more slowly.
- Fractionations in chemical reactions are larger (> 60 ‰) than those in physical processes, because bonds have to be broken.
- Enzymatic reactions show smaller fractionation than the corresponding chemical reaction, because of the multistep mechanisms of many enzymes. Changes in temperature affect the various elementary steps in the reaction differently which may, in turn, cause large changes in isotope fractionation.

Carbon isotope discrimination of plant compounds is commonly given with respect to the limestone standard Pee Dee Selemnite (PDB) as

$$\delta\,[‰] = \frac{R_S - R_{PDB}}{R_{PDB}} \times 1000$$

where R_S and R_{PDB} (=0.0112372) are the molar abundance ratios $^{13}C/^{12}C$ of the sample compound and the standard. The deuterium/hydrogen standard is Vienna Standard Mean Ocean Water ($R_{V-SMOW} = 0.00015576$).

Prior to mass spectrometric analysis the flavor compound is combusted to CO_2 and H_2O. The average CO_2 isotope ratio of the molecule can then be measured directly, while H_2O has to be reduced to H_2. Derivatization of hydroxy groups or other interfering hydrogen bearing functions eliminates undesirable $^2H/^1H$ exchange reactions. Deuterium measurement is indicated if carbon isotope values overlap with synthetic compounds or fall outside the expected range (Isotope Ratio Mass Spectrometry, IRMS; or Stable Isotope Ratio Analysis, SIRA). Samples to be submitted to IRMS are often purified by chromatography. It must be considered that both gas and liquid chromatography, depending on the kind of stationary phase and temperature, may partially or completely separate isotopomers (Braunsdorf et al., 1993). ^2H- and ^{13}C-substituted compounds elute in front of their lighter counterparts during gas chromatography (Berger et al., 1990); the reverse effect occurs in liquid chromatography. Physical concentration, for example using countercurrent chromatography, can also shift $^{13}C/^{12}C$-ratios (Braunsdorf et al., 1993). To reveal discrimination effects during clean-up or analytical processing, classical isotope analysis can be used. The classical technique and GC-IRMS were compared in a study on pineapple volatiles, and almost identical values were found (Schmidt et al., 1993). Correlations of δ-C values and ^{14}C activity were reported for eleven flavor chemicals from different chemical classes (Culp and Noakes, 1992).

Site specific Natural Isotope Fractionation measured by Nuclear Magnetic Resonance (SNIF-NMR) quantifies monodeuterated isotopomeres. Large amounts of pure compound (50–100 mg) are required for this costly method. The specific intramolecular distribution of deuterium in volatile flavors is related to kinetic and thermodynamic isotope effects during biosynthesis and provides a unique fingerprint of the respective origin. Fragments of a molecule are representative of a specific biosynthetic pathway. Volatiles analyzed comprise vanillins, anethols, benzaldehyde, carboxylic acid esters, terpenes, organic acids, and lactones (Fronza et al, 1993; Tateo et al., 1993). Data bases of the site specific isotope ratios are continuously being developed; for example, several thousand samples of ethanol have been collected. ^{13}C-SNIF-NMR may be applied to detect adulteration by adding ^{13}C enriched chemicals, but takes a rather long spectrometer time.

5 Aroma Compounds From Microbial De Novo Synthesis

The aroma compounds of traditionally fermented foods originate, at least in the first phase, from a complex microflora that acts in an only partially understood way on the chemical precursors of a complex food matrix (cf. chapter 2). Most of the classical processes are highly self-regulated and result in constant patterns of volatiles, others depend on narrow windows of the biocatalyst's properties and manufacturing variables with the imminent risk of off-flavor formation. A concerted biotechnology of aromas requires a defined microbiology and nonfood nutrient media. The following chapter will mainly discuss selected work carried out with modified, but usually not optimized media to illustrate scope and depth of current research. From an application point of view a subdivision of topics according to the chemical structures identified was thought to be preferable.

5.1 A Short History of Bioflavors

Besides chemical and morphological criteria, the odor emanated by cultivated microorganisms served to classify species in the nineteenth century. A first comprehensive study on microbial aroma generation appeared as early as in 1923 (Omelianski). The often cited paradigm of *Pseudomonas aeruginosa*, a human pathogen with a sweetish odor caused by 2-amino acetophenone, demonstrates the diagnostic value associated with the odor impression (Mann, 1966). Experienced physicians pay attention to the odor of breath, urine or infected organs. Species with striking odor properties have been labeled with epithetons such as *aromaticus, esterifaciens, odorus, odoratus, fragi,* or *nobilis.* Omelianski qualified aroma formation as a 'changeable character' which was more or less rapidly lost during 'artificial cultivation'. He recommended natural substrates, for example strawberry leaves for regaining strawberry aroma. While these comments may have been the first, attentive observations of glycosidic flavor precursors plants, they have also prepared the ground for the continued use of complex nutrient medium ingredients with their incalculable biochemical consequences. Omelianski correctly noticed that changes of the odor of a microbial culture may arise from the concentration dependent impressions imparted by a single volatile, or from a dynamic change of the entire profile of volatiles during batch cultivation. The isolation of a yeast-like fungus in 1889 is described. The odor of this *Sachsia suaveoleus* was probably caused by methylbutanols and some of their esters. Initially called 'Weinbouquet Schimmelpilz', this strain might have been identical with a strain of *Oidium suaveoleus* or *Oospora suaveoleus* that was, almost one century later, reported as a source of fruity odors (Hattori et al., 1974). Omelianski also reported a monotrichic bacillus isolated from milk and producing a strawberry odor. The ester chemistry of this *Pseudomonas fragi* was fully evaluated only recently (Cormier et al., 1991; Raymond et al., 1991; cf. Sect. 5.3.5). The superior potential of higher fungi was early recognized (Badcock, 1939). In the 1940s the group led by Birkinshaw initiated a series of studies on volatile metabolites of higher fungi (Birkinshaw and Findlay, 1940; Birkinshaw and Morgan, 1950). Without the aid of gas chromatography, volatiles of *Lentinus lepideus* and *Trametes suaveolens* were identified as anisaldehyde and methyl esters of *p*-methoxy cinnamic acid, anisic acid, and cinnamic acid. Species of *Ceratocystis (Endoconidiophora* at that time) formed methylpropyl acetate and methylheptene structures. Volatile, regular terpenes were first detected in *Ceratocystis* in the early 1960s, and later on in other species (Sprecher et al., 1975; Drawert and Barton, 1978). The unique spectra of volatiles of many *Ceratocystis* species have fascinated researchers all over the world (Collins and Morgan, 1962; Collins, 1976; Lanza et al., 1976). Bacterial volatiles received little interest except in cases, where off-flavor problems, for example caused by geosmin in drinking water supplies, were to be tackled (Gerber, 1968; Collins et al., 1971). Food related species, such as *Penicillium roquefortii* (Gehrig and Knight, 1958), or *Candida* (Okui et al., 1963) were cultivated on minimal media to study the formation of fatty acid breakdown volatiles. The microbial portion of cheese flavor and the flavor of edible mushrooms have been elucidated in thorough studies (for example, Dwivedi and Kinsella, 1974; Pyysalo, 1976). With the advent of coupled gas chromatography-mass spectrometry the number of publications and patents increased considerably. Summarizing the state of the art of 1980, Kempler concluded that 'the industrial demand for flavoring compounds .... coupled with elucidation of the biosynthetic pathways involved ... will undoubtedly result in ever-increasing applications of microbial processes for flavor production in the coming years' (Kempler, 1983).

5.2 Oligo – Isoprenoids

Open chain and cyclic mono- and sesquiterpenes with different degrees of oxidation are the chemical and sensory principles of most essential oils. Commercially important monoterpenes, such as D-limonene (citrus), citral (lemon), menthol (cooling), carvone (caraway/spearmint) and α-terpineol (lilac) are abundant in readily available essential oils and are usually purified by rectification. As a result, the plant sources of terpenoids favorably compete with microbial synthesis.

The bacterial synthesis of volatile terpenoids was limited to geosmin, some bornane derivatives, and cadin-4-ene-1-ol (Gerber 1968; Collins et al., 1971; Gerber, 1971). Yeasts as a source of volatile terpenes were discussed in section 2.3.1. *Kloeckera* and *Torulopsis* strains were capable of forming low amounts of limonene, linalool, α-terpineol, and β-myrcene. Free farnesol as a central intermediate to higher terpenoids, appeared to leak from cells of *S. fermentati, S. rosei*, and also *S. cerevisiae*, if culture conditions did not favor regular growth. Citronellol biosynthesis by *Kluyveromyces lactis* did not depend on special precursors, but was stimulated by the addition of L-asparagine (Drawert and Barton, 1978). This preferred role of L-asparagine has, meanwhile, been confirmed for many other microorganisms. Strains of *Ambrosiozyma*, a yeast-like organism with uncertain taxonomy, were the only exception to the rule and produced common monoterpenes, such as geraniol or citronellol, in the mg L^{-1} scale (Klingenberg and Sprecher, 1985).

'Fungi invariably receive a bad press' (Wainwright, 1992), but they are by far the most versatile producers of volatile terpenoids. This included the metabolic main stream compounds geraniol, nerol, and linalool and their sesquiterpene analogs farnesol and nerolidol (Hanssen and Abraham, 1987; Abraham and Berger, 1994), as well as rare structures that have not been identified in any plant source (Abraham and Arfmann, 1988; Fig. 5.1). Fungal cultures need to be aerated. Depending on incubation time and general cultivation conditions, linalool was rapidly transformed into cyclic ethers that have also been found in many essential oils (Abraham and Berger, 1994; Fig. 5.2). Linalool exhibits a very low odor detection threshold, and the cyclic oxides were supposed to contribute to the odor of essential oil due to their sometimes large abundance. Still best researched and unsurpassed in the diversity and quantity of monoterpenes produced are various species of *Ceratocystis* (Sprecher et al., 1975; Collins, 1976; Lanza et al., 1976). The most common terpenoids are listed in Fig. 5.3. Among the terpineols, the α-isomer and its acetate preponderate. Though the rate of uptake of 2-^{14}C-mevalonate was low, it was concluded that the biosynthesis followed the pathway as established in higher plants (Lanza and Palmer, 1977). The usually low yields were increased more successfully by the application of a polymer adsorbent than by modifications of the fermentation medium (Schindler, 1982). The adsorbent, XAD-2, was contained in an external vessel, and the bioreactor broth was circulated through the fixed bed to accumulate the volatiles continuously (cf. chapters 10, 11). About 1.9 g×L^{-1} of odorous metabolites was finally recovered which is close to dry weight related figures of essential oil plants. The previous results can be summarized as follows:

– Ascomycetes (*Ceratocystis*) and a series of basidiomycetes produce mono- and sesquiterpenes in both quality and quantity close to plant sources.

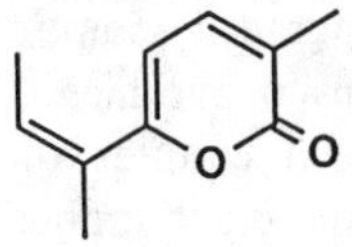

Fig. 5.1. 6-((Z2)-butenyl)-3-methyl-α-pyrone, a perconjugated monoterpene generated by *Fusarium solani* (Abraham and Arfmann, 1988)

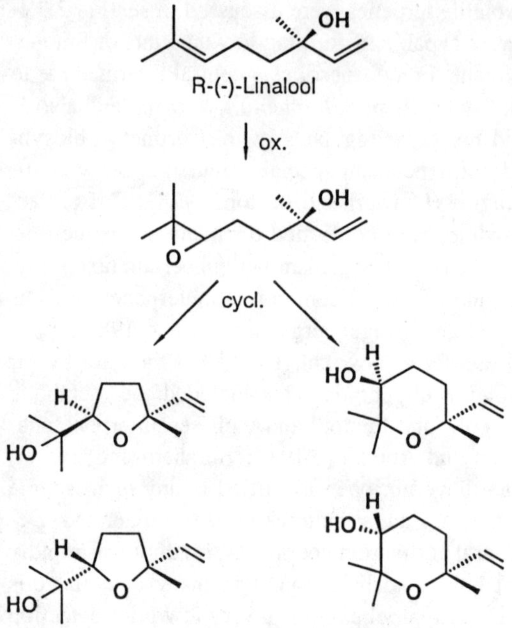

Fig. 5.2. Oxidation of genuine (microbial) linalool results in chemically formed linalool oxides

- Formation of volatile terpenes strictly depends on the strain (or mutant) used, and a production strain must be selected as usual in biotechnology.
- Optimization of the nutrient medium (for example L-asparagin vs nitrate) affects quality and quantity of products more than feeding mevalonate.
- In situ removal of terpenes will be inevitable to receive anything like economically attractive yields.

Major obstacles to higher yields and industrial applications are evident: There is still considerable lack of knowledge on regulation of terpene biosynthesis in microorganisms, and hopanoid research has revealed that even the very first steps of isoprene synthesis in bacteria might not proceed according to the general textbook route. The more powerful monoterpene aldehydes and ketones are rarely found. Some difficult to synthesize sesquiterpene structures, such as α-copaene (Abraham and Berger, 1994), α-bisabolol (Gross et al., 1989), eleganthol (Arnone et al., 1993), or drimene skeletons were identified (Hanssen and Abraham, 1987; Abraham and Berger, 1994), but structurally related flavor and fragrance compounds were not formed (Fig. 5.4). The sesquiterpene hydrocarbons α-gurjunene, (E)-caryophyllene, and a cadinene were isolated from submerged cultures of strains of *Aspergillus flavus* (Zeringue et al., 1993). Four out of the eight strains accumulated and degraded sesquiterpenes and aflatoxins, while the nonaflatoxigenic strains did not produce volatile terpenes. This correlation obviously cannot be explained by a common biogenetic derivation. Similarly, many microorganisms

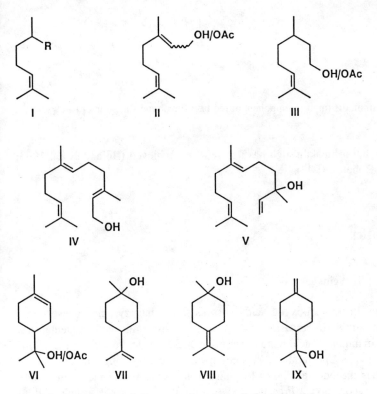

Fig. 5.3. Terpenoids of *Ceratocystis* strains. I=6-Methyl-5-hepten-2-ol and derivatives, R=OH/OAc/O, II=Geraniol/Nerol/acetates, III=Citronellol/acetate, IV=E,E-farnesol, V=E-Nerolidol, VI – IX=α – δ-Terpineol

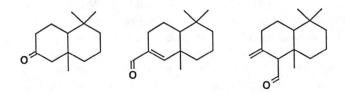

Fig. 5.4. Drim-8(12)-ene derivatives with ambergris attributes structurally related to sesquiterpenes isolated from basidiomycete cultures

that are appreciated for their rich spectrum of secondary metabolites produce the norterpenoid geosmin (Fig. 5.5). This tertiary alcohol is a potent earthy odorant produced by a wide range of *Streptomycetes* (Rezanka et al., 1994), *Penicillia, Basidiomycetes*, and cyanobacteria (Pollak, 1994). Nanogram amounts per liter render water undrinkable; the compound is however, considered valuable in red beet and whisky flavor and in perfumery. With this in mind, attempts have been made to elucidate the biosynthesis of geosmin by either suppressing or stimulating

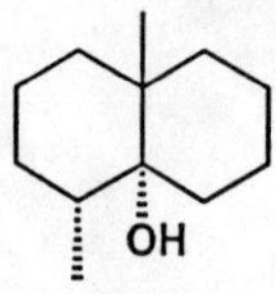

Fig. 5.5. Geosmin, the impact component of red beet flavor and soil streptomycetes

its formation, but a breakthrough has not yet been achieved (Boland et al., 1993; Mattheis and Roberts, 1992).

5.3 Aliphatics

5.3.1 Carboxylic Acids

Medium chain, linear carboxylic acids possess sharp, buttery, and cheese odors that change to fatty with increasing chain length. Methyl branched representatives, such as 3-methylbutanoic acid have lower odor thresholds and originate from amino acid precursors. Unsaturation, depending on the configuration of the double bonds, modifies the odor attributes to pungent, spicy, and with increasing chain length, to fatty and tallowy. Free fatty acids not only occur in the aroma of lipid foods, such as cheese, but also accentuate fruity and fermented beverage aromas. The common, linear fatty acids are formed during repeated β-oxidative cycles upon lipolysis of regular triacylglycerols and are found, consequently, in many of the mixed food fermentations (cf. chapter 2).

Production of dilute acetic acid, the start of modern solid state biotechnology, depends on immobilized cells of *Acetobacter aceti*. Ethanol is more efficiently oxidized by high cell concentrations, as they can be achieved in hollow fiber modules. Immobilization by entrapment of *Propionibacterium* for the generation of propanoic acid was mentioned in chapter 2. High dilution rates during operating the continuous process were applied for high yields and reduced risk of contamination. Butanoic acid has been an undesirable side product in acetone/butanol solvent fermentations using anaerobic *Clostridium butyricum* or *C. acetobutylicum*. With the increasing value of natural butanoic acid and some of its esters the process was reevaluated and the metabolism shifted to butanoic/acetic acid production. A medium pH value near neutral and low nutrient concentrations were essential. High yields of up to 20 g×L^{-1} have been reached in optimized batch fermentation. The inhibitory effects of the acids on their producer cells have been recognized and are thought to be eliminated by coupling the bioprocess with advanced downstreaming procedures (cf. chapter 10). The same toxic situation likewise limits the accumulation of carboxylic acids by yeasts. Higher fungi are usually more acid tolerant and can even grow at pH 2 and below (Hädrich-Meyer and Berger, 1994). Additionally, yeasts, other fungi, and plant cells have evoluted rapid detoxification mechanisms, such as reduction and esterification. The rational approach of using food strains and food imitating, simplified media has led to fatty-acid-based cheese flavors (Jollivet et al., 1992). Strains of *Brevibacterium* and a *Microbacterium* strain produced volatile fatty acids, alcohols, methylketones, thio compounds, and aromatics when cultivated in a tryptic soya medium. The aroma profiles of the four strains of *Brevibacterium* were similar and coined by dimethyl trisulfide. The floral compounds 2-phenyl ethanol and 3-phenyl propanol occurred in significant amounts in three of the strains. An emulsion of water or milk in butterfat was converted to a cheese-like aroma carrier by using *Penicillium camemberti*, *P. roquefortii*, or *Aspergillus niger* (Lemenager et al., 1992). A number of similar patents in the field depended on the same microbiology.

5.3.2   Lipoxygenase and Cyclooxygenase Products

Fungal lipoxygenases (EC 1.13.11.12) stereospecifically catalyze the peroxidation of linoleic and linolenic acid to 10-monohydroperoxides. The volatile impacts of mushroom aroma are generated by the activity of a hydroperoxide lyase and further steps (Fig. 5.6). Several suggestions have been made regarding the mechanism of the fragmentation reaction, but further work is required to fully characterize the entire sequence. Physical or biological attacks that damage the integrity of the fungal cell start the reaction by bringing together lipoxygenase and its substrate. A comprehensive study has recently confirmed the impact role of the eight carbon carbonyls (Buchbauer et al., 1993). Headspace sampling, steam distillation, and solvent extraction were compared to isolate aroma compounds from *Agaricus campestris*, *A. bisporus*, *Lepiota procera*, *Armillaria mellea*, *Boletus edulis*, and *Cantharellus cibarius*. The lipoxygenase derived carbonyls were not only identified in the comminuted 'tissues', but also in the headspace above intact fruiting bodies. This unexpected result may be explained by considering the impossibility of harvesting (and transporting) the fungal materials without disrupting viable cells. Some esters and aromatic aldehydes were contributory and accumulated in the distillation and extraction samples. Agricultural treatments have been investigated to improve lipoxygenase activity (Mau et al., 1993). Immature mushrooms with closed veils had higher octen-3-ol levels than mature mushrooms with open veils. Storage temperature affected the activity of the hydroperoxide lyase without changing octen-3-ol contents. Addition of calcium chloride to the irrigation water increased the amounts of this alcohol after harvest. More octen-3-ol was produced in the gills of the fruiting body than in other parts of the mushroom. It is a remarkable coincidence that gills of rainbow trout were the preferred location for an animal lipoxygenase (German et al., 1991). In both cases the lipoxygenase bearing tissue is the one most exposed to the surrounding atmosphere. The second product of the hydroperoxide lyase, 10-oxo-(E8)-decenoic acid (Fig. 5.6), stimulated mycelial growth and stipe elongation in *A. bisporus* (Mau et al., 1992). This agrees with the anticipated involvement of lipoxygenase products in the chemical defense system (cf. Sect. 1.5). The same acid has also been suggested to initiate fruiting.

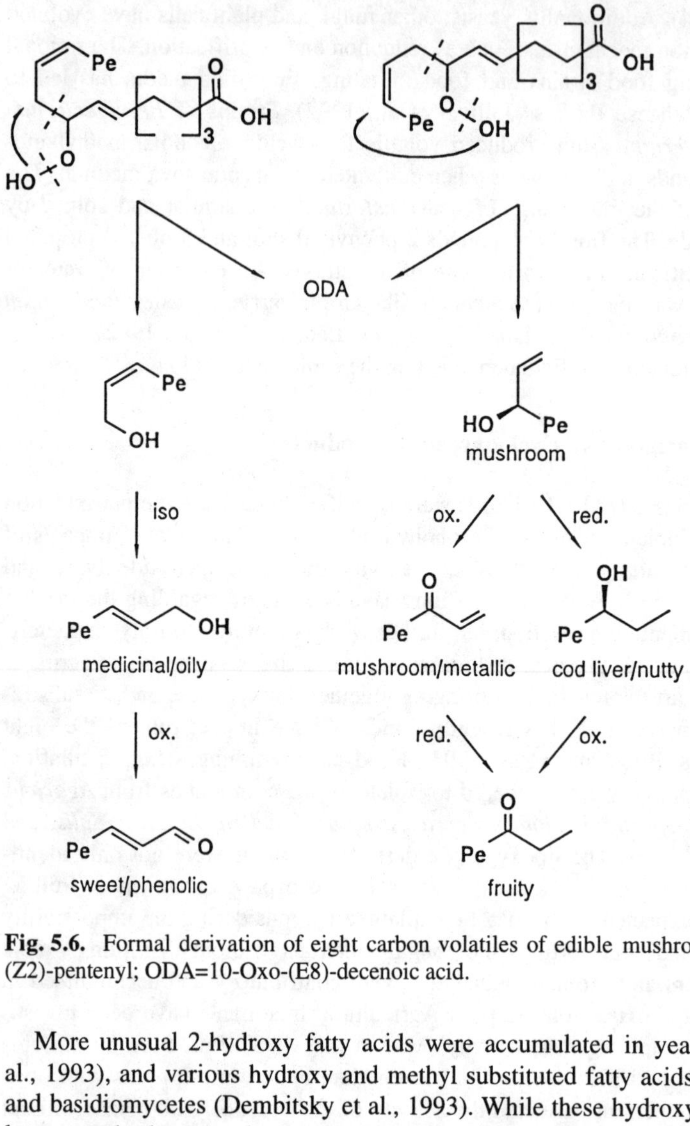

Fig. 5.6. Formal derivation of eight carbon volatiles of edible mushroom; Pe=pentyl or (Z2)-pentenyl; ODA=10-Oxo-(E8)-decenoic acid.

More unusual 2-hydroxy fatty acids were accumulated in yeast (Kaneshiro et al., 1993), and various hydroxy and methyl substituted fatty acids in ascomycetes and basidiomycetes (Dembitsky et al., 1993). While these hydroxy acids are odorless as such, they may be degraded by microbial activities to novel, volatile cleavage products.

Jasmonic acid and derivatives in microorganisms may originate, as supposed for plants, from the same eighteen carbon precursor fatty acid via (Z9, E11, Z15)-13-hydroperoxyoctadecatrienoic acid (Vick and Zimmermann, 1984). A hydroperoxycyclase delivers in a prostaglandin analog synthesis a cyclopentanoyl carboxylic acid that, upon 3 fold β-oxidation, yields the precursor acid of the odorous methyl epijasmonate (Fig. 5.7). Some of the jasmonoid compounds regulate plant growth and senescence. In view of the ubiquitous precursor it is not surprising that jasmonates were found in many plant sources. The methyl ester of epijasmonic

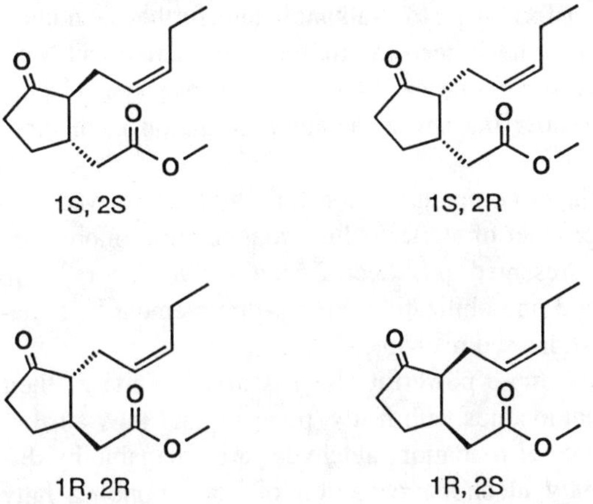

Fig. 5.7. Stereoisomer methyl jasmonates: 1S, 2S, and 1S, 2R are odorless; 1R, 2R is weak; 1R, 2S strong, threshold 3 μg L^{-1}

acid contributes to the precious flavor of jasmin oil, and also to *Rosmarinus, Gardenia, Arthemisia*, and lemon peel oil and to black tea aroma. Derivatives and conjugates of jasmonic acid were isolated from culture filtrate of a mutant strain of *Gibberella fujikuroi* and from a strain of *Botryodiplodia* (syn. *Lasiodiplodia*) *theobromae* (Miersch et al., 1993). The ecological function of jasmonic acid in microbial cultures is still unknown (cf. chapter 1, Table 1.2).

5.3.3 Carbonyls

Among the ketones, odd-numbered 2-alkanones from five to eleven carbons, along with free fatty acids and 2-alkanols, determine the aroma of *Penicillium* ripened cheese and have received much attention. The biogenesis results from an overflow of the β-oxidation cycle, where an excess of 3-ketoacyl-CoA ester is accumulated. A part of this pool is not processed further to acetyl-CoA and acyl-CoA, but spontaneously hydrolyzed followed by decarboxylation. For blue-cheese flavor production, conidiospores or mycelium of *Penicillium* are suspended or immobilized and contacted with dairy waste streams supplemented with coconut oil or medium chain free fatty acids. Continuous and fed-batch systems with integrated product recovery reflect the advanced state of research in this field (van der Schaft et al., 1992a). Bacteria, such as *Aureobasidium* (Fukui and Yagi, 1991), yeasts, and higher fungi produce 2-alkanones, but only *Penicillium* is used industrially. A comparative study of ten strains of *P. camemberti* in a cheese related medium found similar spectra of volatiles, but significant quantitative differences (Jollivet et al., 1993). The best strain produced more than 50 mg of 2-alkanones×L^{-1} after 4 days of incubation. More than 60% conversion of added fatty acid precursor have been reported. After 14 days of incubation the concentrations of 2-alkanones and

total volatiles had decreased. Transformation to 2-alkanols and further to nonvolatile hydroxy ketones and diols may have been the reason. Alternative pathways
to 2-alkanones, such as alkane oxidation or CoA-independent conversion of fatty
acids are under discussion and render the entire metabolic situation even more
difficult.

The only important diketone flavor compound is the carbohydrate derived 2,3-
butandione. Available as a side product of starter culture manufacture, more concerted bioprocesses have been presented (cf. Sect. 2.5). Crucial factors were
supply of citrate, water activity, and immobilization. Strain improvement by mutagenesis or genetic engineering is addressed in chapter 8.

Aliphatic alkanals and the often more powerful alkenals are, in spite of their
chemical lability, valuable odor compounds with fruity, pungent, and fatty attributes. Through their intermediate state of oxidation, aldehydes are amenable by different routes: oxidation of primary alcohols, reduction of corresponding fatty
acids, aldolase reaction, or by a STRECKER analog decarboxylation of 2-keto
acids. Acetaldehyde, in dilution an etheral top note in citrus and yoghurt flavors, is
a good example for a structurally simple, but commercially important molecule
that has become a biotechnological target compound. Market demands have stimulated research by a Canadian group to convert (bio)ethanol with adapted *Candida
utilis* or alginate immobilized *Pichia pastoris* cells (Armstrong et al., 1993).
Though the low boiling point of acetaldehyde (21 °C) would suggest a gas stripping low-temperature trapping system, a TRIS buffer was used to trap the product
in situ. Thus, the problem of end-product inhibition was reduced, but the 'natural'
character of the aldehyde remained arguable. Highly aerated mutants of *Zymomonas mobilis* can produce acetaldehyde from glucose. Basically the same problems
of cytotoxicity of the product were encountered. Enzyme technology using alcohol
dehydrogenase removes all these drawbacks and creates the new problem of cofactor regeneration (cf. Sect. 7.6). Another novel approach was a gas/solid bioreactor
filled with dried whole cells of *Hansenula polymorpha* (Kim and Rhee, 1992).
This reactor was operated for over one month at 35 °C with a complete conversion
of ethanol. Due to the low substrate specificity of alcohol oxidizing microbial
enzymes other than aliphatic alcohols were also accepted, for example benzyl
alcohol and cinnamyl alcohol (Armstrong et al., 1993). A long chain aldehyde
forming enzyme activity was reported for the green alga *Ulva pertusa* (Kajiwara et
al., 1993). Products with a seaweed flavor were (Z8, Z11, Z14)-heptadecatrienal
and others. The origin from unsaturated fatty acids is again obvious. No such
activity has yet been detected in microbial cells.

5.3.4 Alkanols

The predominant volatiles of all fermented beverages are, in addition to ethanol,
the 'fusel oil' alkanols 2-methylpropanol, 2/3-methyl butanol and 2-phenyl ethanol.
They are formed by most yeasts (cf. Sect. 2.3.1) either during synthesis or degradation of amino acids via 2-keto acid intermediates. Straight-chain even-numbered
alkanols and the alkyl moieties of the respective carboxylic acid esters of fruits are

lipid catabolites generated by plant reductases. Similar bacterial reductive conversions by *Clostridium* species can afford alkanols from fatty acid precursors as variants of the acetone/butanol fermentation. The extraction solvent can be used as a reaction medium for the in situ esterification by microaqueous lipases (cf. Sect. 7.1.3). The same reaction sequence must be assumed in ester forming microorganisms, such as *Streptococcus lactis* and *Pseudomonas*.

More or less pronounced, all of these pathways to higher alkanols operate in higher fungi, the methylbutanols and 2-phenyl ethanol being almost ubiquitous volatiles. The eight carbon saturated and unsaturated (Fig. 5.6) alcohols of the lipoxygenase catalyzed lipid degradation characterize aging or mechanically/chemically mistreated fungal cultures. The above mentioned and many other alkanols were identified in submerged cultured strains of the basidiomycete *Phlebia radiata* (Gross et al., 1989; Fig. 5.8). Alcohols, including phenyl substituted compounds, regularly constitute the most numerous group of volatiles of basidiomycete cultures (Gallois et al., 1990). Yeasts and filamentous fungi with relevance in human mycology produced fusel alcohols and volatile esters, while keratinolytic fungi were much less active (Cailleux et al., 1992). The cutaneous flora appears to be more bacterial. A lipophilic *Corynebacterium*, when incubated with pooled apocrine secretion, transformed an unknown, possibly steroidal precursor to (E3) and (Z3)-methyl-2-hexenoic acid (Zeng et al., 1992). This work was directed at identifying the characteristic human axillary odors and their role in human chemical communication (cf. Table 1.1). The presence of 2-ethylhexanol was correlated with a pungent odor that occurred in buildings affected by contamination with *Aspergillus versicolor* (Bjurman et al., 1992).

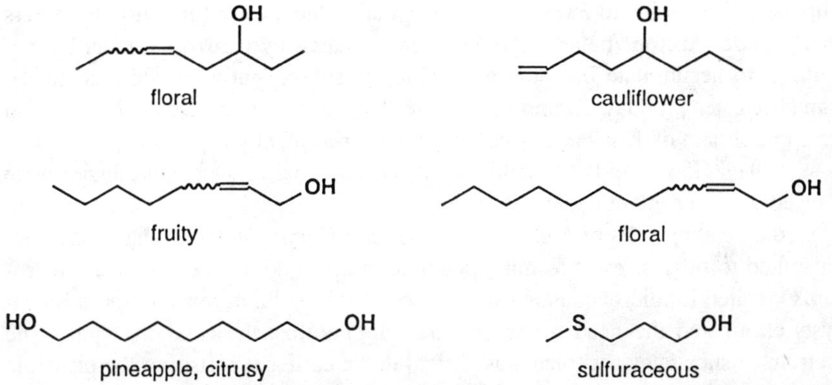

Fig. 5.8. Volatile alkanols (stereochemistry unknown) of *Phlebia radiata* (Gross et al., 1989)

5.3.5 Esters

Aliphatic esters are key compounds in fruit flavors, but also contribute to the overall aroma of chocolate and fermented beverages. Of particular commercial importance are ethyl propanoate, butanoate and laurate, and methylbutyl acetate. As discussed above, acyl and alkyl moieties can be prepared by many microorganisms along common pathways (cf. Sect. 2.3.2). The alkyl moieties are acylated by enzymes that transfer CoA-activated fatty acid moieties. The absolute amounts of the single ester compounds depend on concentration and availability of the substrates, and on the specificity of the transferase. Accordingly, ethyl acetate and further ethyl esters dominated in unsupplemented yeast cultures (Davies et al., 1951). The esters formed by *Hansenula mraki* were stripped-off by an air stream and adsorbed in charcoal traps (Janssens et al., 1988). The yeast cultures have to be cultivated with essential nutrients, such as unsaturated fatty acids, and under suitable partial pressures of oxygen and carbon dioxide. However, active growth must be controlled, for example by metal limitation, to avoid a depletion of the acetyl-CoA pool. In agitated liquid media that simply contained glycerol and an amide *Geotrichum candidum* produced a quince/pineapple aroma (Latrasse et al., 1987). *G. candidum* (formerly *Oospora lactis*, a relative of the 'Weinbouquet Schimmelpilz') is a yeast-like fungus that occurs on cheese. Among the volatiles identified were seven ethyl esters, a series of methylpropyl esters, and the fusel oil alcohols. The fruity character was attributed to the mixture of ethyl methylpropanoate and methylpropyl acetate. Isoprenoid derived alkyl moieties of fungal esters were described above (Fig. 5.3).

Bacterial volatile esters were first met in off-flavored dairy products. Several genera of lactic acid bacteria and of *Pseudomonas* generated ethyl butanoate, ethyl 3-methylbutanoate, and ethyl hexanoate in milk-like chemical environments (Cormier et al., 1991; Raymond et al., 1991). About 90 microbial volatiles were isolated from a skim milk, whey, or whey permeate based nutrient media. Addition of lipids or fatty acids to *Pseudomonas fragi* stimulated ester formation by orders of magnitude. Acetone/butanol fermentation on starch by *Clostridium* can be manipulated to accumulate butanoic acid which is subsequently transformed to the symmetric ester product. Amino acids were the precursors of aroma alcohols and esters in cultures of *Erwinia carotovora*, a bacterial plant pathogen (Spinnler and Djian, 1991). Good yields of methionol, its acetate, and of 2-phenyl ethanol were achieved using a synthetic nutrient medium.

Dried or resting cells of *Mucor, Rhizopus*, and *Corynebacterium* have been demonstrated to possess ester forming potential under conditions of reversed hydrolysis. Operated in microaqueous batch or continuous column reactors these biocatalysts eliminated the need to isolate and reimmobilize the catalytic agent. The spectrum of substrates accepted was limited in the earlier reports, but the principle question remains if the catalysis of a single step reaction will really proceed better with a purified enzyme (cf. Sect. 7.1.4).

5.4 Aromatics

The aromatic flavors consumed most are methyl salicylate and anthranilate, and benzaldehyde, cinnamaldehyde, and vanillin. These and many other aromatic

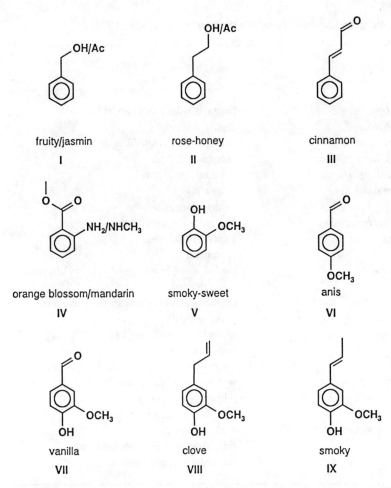

Fig. 5.9. Aromatic impact components. I = Benzyl alcohol/acetate, II = 2-Phenyl alcohol/acetate, III = Cinnamaldehyde, IV = Methyl/*N*-methyl anthranilate, V = Guaiacol, VI = Anisaldehyde, VII = Vanillin, VIII = Eugenol, IX = Isoeugenol

chemicals can determine the flavor of foods and essential oils. All of the volatiles shown in Fig. 5.9 have been identified in microbial sources. Except for 2-phenyl ethanol and its derivatives that are ubiquitous, volatile aromatics are the domain of filamentous fungi (Table 5.1).

The preferred metabolism of aromatic compounds results from the ecological niche occupied by the white-rot fungi: dead or living wood. Lignin is a random, peroxidase catalyzed polymer of substituted *p*-hydroxy-cinnamyl alcohols, and white-rot fungi are the only organisms able to degrade this polymer network totally. Most white-rot fungi belong to the *Basidiomycetes* group, some are *Ascomycetes* (for example, *Xylaria hypoxylon*). Some *Actinomycetes* represent bacterial lignin degraders, but they are less efficient and have not yet been reported to gene-

Table 5.1. Fungi are the main source of volatile aromatics (Tiefel, 1994)

Species	Product(s)
Agaricus bisporus etc	benzaldehyde, phenyl acetaldehyde
Ascoidea hylecoeti	2-phenyl ethanol
Bjerkandera adusta	anisaldehyde, veratraldehyde
Camarophyllus virgineus	anisaldehyde
Hebeloma sacchariolens	2-amino benzaldehyde
Hyanellum suaueolens	coumarins
Inocybe sp.	methyl cinnamate
Ischnoderma benzoinum	benzaldehyde, anisaldehyde
Lentinus sp.	benzyl acetate, methyl anisate, methyl cinnamate
Hycoacia vda	methyl acetophenone, methyl benzylalcohol, *p*-tolualdehyde
Nidula sp.	raspberry ketone, cinnamic acid derivatives
Phanerochaete chrysosporium	veratraldehyde
Phellinus sp.	methyl benzoate, salicylate
Pleurotus euosmus	coumarin
Poria sp.	anthranilic and cinnamic acid derivatives
Pycnoporus cinnabarius	methyl anthranilate, vanillin
Sirodesmium diversum	*p*-hydroxybenzaldehyde
Sparassis ramosa	oak moss impacts (orcine derivatives)
Stereum subpilatum	methyl coumarate
Trametes sp.	anisaldehyde, methyl phenylacetate
Tyromyces sambuceus	benzaldehyde, ethyl benzoate

rate aromatic volatiles. Lignin peroxidase, a manganese dependent peroxidase, and laccase constitute the lignolytic system. Several hundred white-rot species are known to the plant pathologist, and many details of lignin degradation were investigated using *Phanerochaete chrysosporium* as a model fungus (Fiechter, 1991). Intra- and extracellular generation of hydrogen peroxide, the peroxidase cosubstrate, has been demonstrated. The primary attack on native lignin and subsequent enzymatic steps are still being discussed controversially, because neither continuous cultivation nor repetitive batch cultures of *Ph. chrysosporum* allowed a stable synthesis of ligninase or total protein to be excreted. This regrettable situation has not impeded attempts to establish a fungal biotechnology of flavors. Important points are:

– Many higher fungi possess a potential for the formation of volatile flavors comparable to specialized parts of seed plants. A significant fraction of the volatiles (one third in a recent survey, Gallois et al., 1990) are aromatic.

- Higher fungi can be grown on simple, synthetic nutrient media. Lignin can be offered as a hydrolysate or in the form of more or less degraded structural units. *Lentinus lepideus* (a brown-rot fungus, cf. Table 5.1) was able to grow on catechin or tannic acid as a sole carbon source (Collett, 1992). Even xenobiotics may be utilized (Cerniglia et al., 1992). Numerous inducable exoenzymes eliminate the need to thermally digest polymer substrates. According to the metabolic flexibility of a saprophytic organism, a change of substrate will usually redirect the pathways leading to volatile compounds (Tiefel and Berger, 1993a).
- Fungal cells were routinely cultivated on agar plates, resulting in growth periods measured in months. Using modern submerged cultivation this drawback is overcome. The growth of *Pleurotus sapidus* was stimulated by the technique of inoculum preparation rather than by size or age of the inoculum (Abraham et al., 1994). A hundred fold dry mass was recovered from the medium after one week of growth.
- A wealth of catalytic properties awaits application in flavor chemistry, for example acetylene/allene-isomerases in norisoprenoid synthesis (Arnone et al., 1992), or metabolites from veratryl alcohol with oxoenol structure (Tuor et al., 1993).

The complex spectra of volatiles may create problems if the biosynthesis of a single character impact component is aimed at (Kawabe and Morita, 1993). Feeding of appropriate precursors is one approach to direct the metabolic events to the accumulation of a single volatile (Krings, 1994). For example, L-phenylalanine was converted to benzaldehyde by *Ischnoderma benzoinum* in high yields, and almost no volatile side products were found under the respective cultivation conditions. This bioprocess is of commercial interest, because the yields from emulsions of seed pits are low, and recovery and purification are difficult (equimolar amounts of hydrogen cyanide are formed concurrently). The *Ischnoderma* cells needed to be incubated under a precisely controlled pO_2 to maintain high levels of benzaldehyde. Natural L-phenylalanine is required to pursue this kind of study further.

Methyl anthranilate, the impact of American Concord grape (Fig. 5.9), was derived from natural methyl *N*-methyl anthranilate that is abundant in mandarin leaf oil. *Trametes* and *Polyporus* strains with high demethylation activity were used, and the methyl ester was recovered by a one step solvent extraction. *N*-formyl methyl anthranilate is formed as a side product, but can be converted by thermal decarboxylation to the target compound (Page and Farbood, 1989). The biogenesis of more complex flavor chemicals will, of course, be more difficult to control and require much more effort for metabolic fine-tuning.

5.4.1 Vanillin

Vanillin, the most universally appreciated aroma chemical and a prime example of a high-value phytoproduct, has in the past been biotechnologically produced by

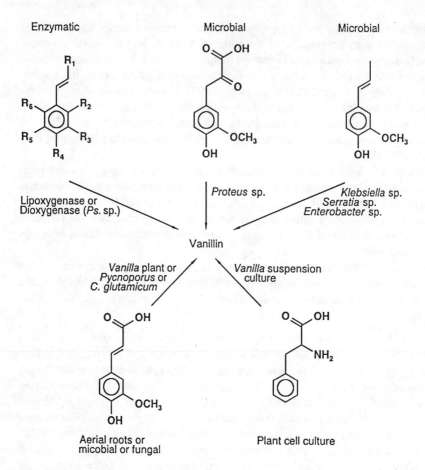

Fig. 5.10. Biotechnology of vanillin (refs. cf. text)

enzymes, microorganisms and plant cultures (Fig. 5.10). Vanillin occurs in the vanilla bean at a level of about a two percent by weight, is associated with many other aroma compounds depending on location of harvest, and is priced at 4000 US$ per kg. The pure chemical of synthetic origin costs about 12 US$ per kg, but only represents the sensory backbone of vanilla. Of the about 12 000 metric tons consumed annually, about 20 metric tons were extracted from beans. The *Vanilla planifolia* qualities grown in Madagascar, Réunion (Bourbon), Mexico, and Indonesia, and the *Vanilla tahitensis* of Tahiti, with significant amounts of heliotropine and anisaldehyde (Fig. 5.9), can be easily distinguished by their specific odor profiles. These differences of price and quality have created a lot of fanciful speculation about alternative processes.

It has occasionally been found that GRAS microorganisms, such as some lactic acid bacteria, increased the concentration of vanillin in food that contained natural vanilla extract or vanilla powder. As the vanilla bean contains vanillin in βglucosidic form (cf. Table 7.2), and water/ethanol mixtures are used as extraction solvents, one may argue that the 'production' of vanillin was nothing more than a mere hydrolytic cleavage of the conjugate. A dioxygenase, isolated from a *Pseudomonas* strain, acted on the double bond conjugated to the benzene ring of styrenes (Hasegawa, 1990). The enzyme accepted a broad range of ring substituents R_2 to R_6, and R_1 was straight or branched, one to ten carbon alkyl, one to six carbon aryl, five to ten carbon cycloalkyl or a carboxy derivative. The enzyme was characterized after several purification steps and claimed to cleave *p*-coumaryl alcohol, coniferyl alcohol, and some further substrates to yield vanillin. A likewise broad spectrum of phenylpropanoids was accepted by a plant lipoxygenase (Quest, 1991). The respective phenyl aldehydes were described in the patent, and vanillin was obtained in moderate (4–15%) yields using a commercial lipoxygenase. Other microbial and plant sources of lipoxygenases, such as *Saccharomyces cerevisiae* and potato juice, were mentioned in the examples given. The use of an enzyme typically related to lipid catabolism for the oxidative degradation of phenolics appears strange, but confirmation of the results by a third party has been achieved. *Proteus vulgaris* converted methoxy tyrosine to the corresponding phenyl pyruvic acid (Casey and Dobb, 1992). The acid was then transformed to the target aldehyde with mild alkali. A considerable range of microorganisms performed the starting reaction. High yields were obtained in a simple process. The drawbacks of this approach were the availability of natural 3-methoxy tyrosine, and the final chemical step. A series of bacteria and fungi were able to convert isoeugenol or ferulic acid to vanillin (Abraham et al., 1988). A patented process was developed from these screenings (Haarmann & Reimer, 1991). Several species of *Serratia*, *Enterobacter*, and *Klebsiella* were claimed. While isoeugenol with its 1,4-conjugated system of double bonds was converted in high yields, eugenol was not. Unfortunately, natural isoeugenol is not yet readily available. The biochemical parallels to lipoxygenase catalysis are obvious. This approach suffered also from insufficient precursor solubility and from oxidation/disproportion reactions of the aldehyde formed. A whole spectrum of side products, such as vinyl guaiacol, phenyl propanoic acid, heliotropin, and veratraldehyde was found, when *Corynebacterium glutamicum* was incubated in the presence of ferulic acid (Labuda et al., 1993). A water soluble sulfhydryl compound was essential in increasing the yield to 15% (of 0.1% ferulic acid precursor) in a ten day fermentation. No protection for this bioprocess has been acquired. Mutants of *Pycnoporus* sp., a basidiomycete, degraded natural ferulic acid to vanillin without significant further conversion of the target aldehyde (Pernod-Ricard, 1991). A typical biosystem contained *P. cinnabarinus* and a sporulation medium with yeast and malt extract, peptone, glucose, biotine, and mineral salts. Low yields of side products were observed with six days old, stationary phase cells. As much as 90 mg vanillin per liter were recovered from the broth by solvent extraction.

Several groups have investigated vanillin formation in plant cell culture. The productivity was enhanced by selecting phytohormones and elicitors, and by immobilization and adsorptive vanillin recovery. Precursors, such as ferulic acid, but also phenyl alanine, dehydroshikimic acid, and vanillyl alcohol proved suitable.

Cell vitality stabilizers favorably supplemented the elicitor treated cultures. The elicitor approach is making use of the chemical defense mechanisms of plant cells. Biotic (compounds of bacterial or fungal origins) or abiotic stress can induce the formation of secondary plant metabolites, among them volatiles (cf. Sects. 1.5, 9.3.4). The compounds newly formed are not necessarily regular constituents of the respective plant species; for example, suspension cultures of *Apium graveolens* produced about 10 mg vanillin×kg⁻¹ dry matter under light stress (Berger, unpublished results). Elicitors may offer an opportunity to increase the concentration of aroma chemicals in low-yielding plant cell cultures.

All biotechnological processes more or less share the same problems:

- Lack of a de novo synthesis that renders the processes dependent of a natural precursor substrate,
- lack of a selective product removal, and a
- lack of biogenetic information prevents an efficient control of side-product formation and of further reactions of vanillin.

A striking amount of work was carried out in the development of vanillin producing bioprocesses, though there is still no clear picture on the metabolic details of the various cellular producers. In suspended cells of *Daucus carota* treatment with fungal elicitors led to the synthesis of *p*-hydroxybenzoic acid. Precursor studies showed that the side-chain degradation from *p*-coumaric acid to *p*-hydroxybenzoic acid was not analogous to the β-oxidation of fatty acids, but involved the formation of the (vanillin related) *p*-hydroxybenzaldehyde as an intermediate. In *Vanilla planifolia* cell cultures a pathway was postulated from 3,4-dimethoxycinnamic acid (not ferulic acid!) via dimethoxybenzoic acid to vanillic acid. This suggestion was supported by the recovery of the radiocarbon label in vanillic acid after supplementation of ¹⁴C-cinnamate, but not ¹⁴C-ferulate.

5.4.2 Cinnamates

Among the cinnamic acid derived volatiles cinnamaldehyde (cinnamon), benzaldehyde (bitter almond, cherry), benzyl acetate (jasmine), methyl benzoate (dry-fruity), and methyl cinnamate (exotic fruit, balsamic) command a significant commercial potential (Fig. 5.11). Many of these aromatics are synthesized in different amounts by *Basidiomycetes* in the absence of a lignin substrate. Supplementation of the nutrient medium with intermediates of the shikimate pathway usually led to volatile transformation products that dominated the volatile fraction. If a precursor substrate, such as phenylalanine or cinnamic acid occupies a central metabolic position, a series of biochemically related volatiles can be formed. Upon feeding L-phenylalanine or L-tyrosin to submerged cultured *Ischnoderma benzoinum*, an immediate formation of benzaldehyde or *p*-hydroxybenzaldehyde was observed (Berger et al., 1987). The addition of castor oil to a *Polyporus* sp. unexpectedly rose the concentrations of benzoates, 2-phenylacetates, cinnamates, and cinnamaldehyde, while a lignin preparation stimulated the formation of eugenol, guaiacol,

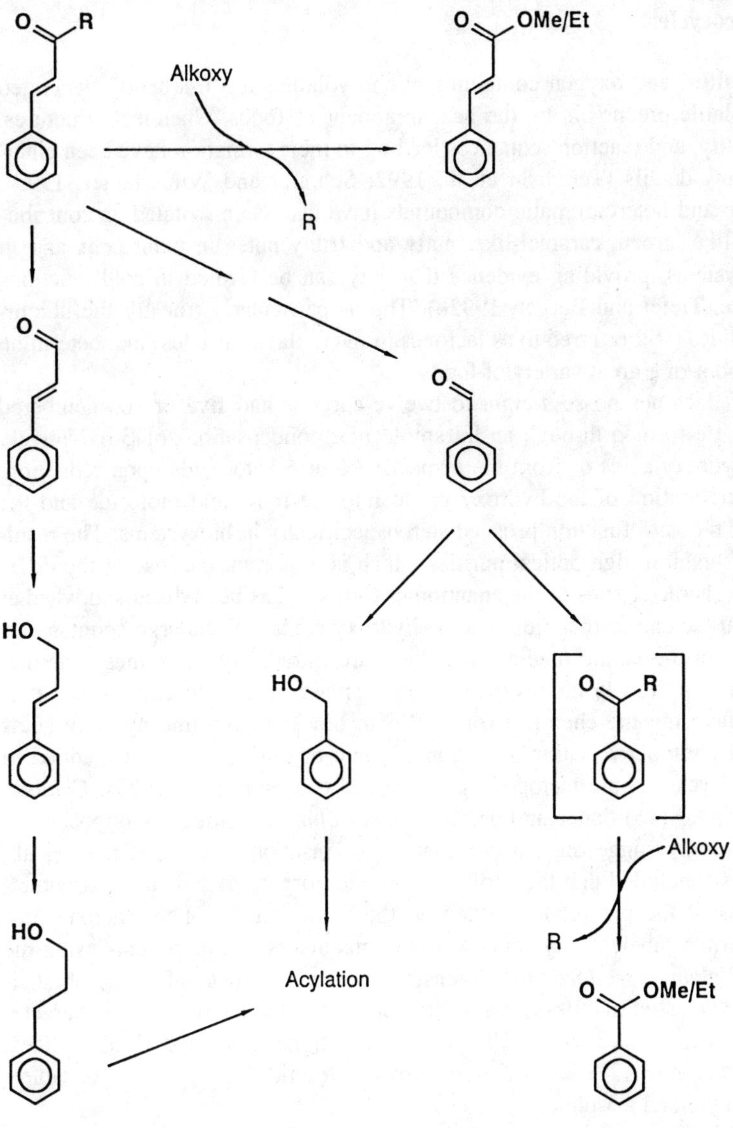

Fig. 5.11. Cinnamic acid derived volatiles of basidiomycete cultures, R = S-CoA (Tiefel, 1994)

isoeugenols, and vanillin (Tiefel and Berger, 1993a). Concentrations ranged from a few to several hundred mg per liter, opening scope for extending the number of accessible bioflavors. In conclusion, many *Basidiomycetes* possess active enzymes for handling aromatic metabolites and should, therefore, be preferably considered for aromatic target volatiles.

5.4.3 Heterocycles

Nitrogen, sulfur, and oxygen containing cyclic volatiles are frequently generated from nonvolatile precursors by the heat treatment of foods. Chemical structures, sensory quality, and reaction sequences leading to their formation have been elucidated in many details (Teranishi et al., 1992; Schreier and Winterhalter, 1993). Heterocyclic and heteroaromatic compounds have also been isolated as contributors of pea-like, green, caramel-like, nutty and fruity notes in plant cells and in microbial systems, providing evidence that they can be formed in cold reactions (for example, Tiefel and Berger, 1993b). This is particularly true for the ubiquitous alkanolides (still referred to as lactones in many flavor articles) that determine the typical odor of a great variety of foods.

Important lactones possess eight to twelve carbons and five or six-membered rings. They are formed through an intramolecular condensation of β-oxidatively shortened hydroxy acids or from the respective 4 or 5-keto acids upon reduction. Both the introduction of the hydroxy group into the fatty acid molecule and the reduction of the keto function proceed stereospecifically in biosystems. The resulting lactones exhibit high optical purities which is important because of the different sensory characteristics of the enantiomer forms. It has been discussed whether the microbial lactone formers just release hydroxy acids that undergo spontaneous lactonization in the acidic media, or if there are intracellular enzymes. Thermal treatment of cell free broth resulted in an increase of the concentration of 4-decanolide favoring the chemical route. When, however, racemic hydroxy acids obtained via photoperoxidation/reduction of oleic or linoleic acid were added to cultured wild yeasts, pure lactones were received (Fuganti et al., 1993). *Candida lipolytica* formed (4S)-dodecanolide, and *Pichia ohmeri* yielded the opposite enantiomer, strongly suggesting an enzymatic condensation reaction (Ercoli et al., 1992). It was concluded that the choice of the microorganism and subtle structural modifications of the precursors influenced the configuration of the lactone products. Deuterium labeled fatty acids were the precursors in experiments using the yeast *Sporobolomyces odorus* to demonstrate the involvement of a microbial lipoxygenase (Tressl et al., 1993). Optically active, labeled lactones were characterized (Fig. 5.12). While 9- or 10-hydroperoxidations generated 4-alkanolides, hydroperoxidation in 13-position followed by heterolytic cleavage of the unsaturated fatty acid yielded a 5-olide.

Microbial lipoxygenases are less characterized and inhibited by higher substrate levels. Commercial bioprocesses for the generation of lactones depend on (plant) hydroxy acids (Table 5.2). Two decades after Okui described the conversion of

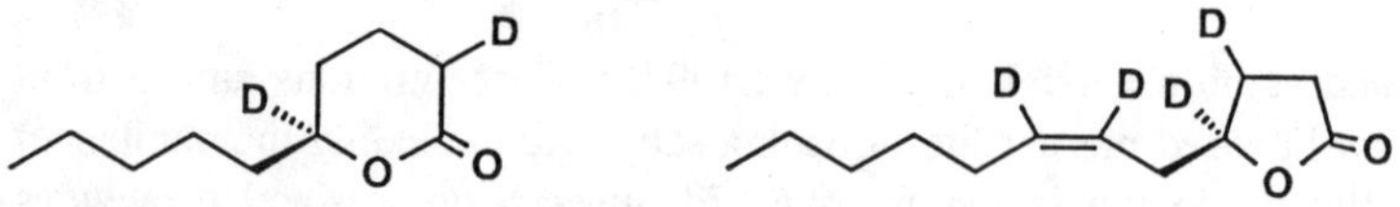

Fig. 5.12. Lactones produced from deuterated fatty acid precursors by *Sporobolomyces odorus* (Tressl et al., 1993).

Table 5.2. Alkanolides from Yeasts, Ascomycetes and Basidiomycetes (Berger, 1991)

Biocatalyst	Product(s)	Year[a]
Trichoderma viride	6-pentyl-2-pyrone	1972
Sporobolomyces odorus	4-decanolide	1972 P
Pityrosporum sp.	4-alkanolides, C_6-C_{12}	1979/87 P
A. oryzae, Geotrichum		
klebahnii, Candida lipolytica	4-decanolide	1983 P
Polyporus durus	4-alkanolides	1983
Fusarium poae	4-alkanolides	1983
Poria aurea	4-alkanolides	1986
Ischnoderma benzoinum	4-alkan/alkanolides	1987
Monilia fructicola	4-and 5-alkanolides	1988 P
Trichoderma, Ceratocystis	5-alkanolides	1988 P
Tyromyces sambuceus	4-decanolide	1989
Mucor	4-and 5-alkanolides	1989 P
Candida sp.	decenolides	1989 P
Cladosporium suaveolus	5-decanolide	1990 P

[a] P indicates a patented process

ricinoleic acid (12-hydroxy oleic acid) to 4-hydroxydecanoic acid by *Candida* yeasts, virtually the same process was patented using castor oil (ca. 90% w/v ricin oleic acid) as the sole carbon source (Fritzsche Dodge & Olcott, 1983). In the following years the high price of biotechnologically produced (4R)-(+)-decanolide (up to 20 000 US$ per kg) started a race for other biocatalysts and precursor acids (Table 5.2). Baker's yeast, for example, can convert 11-hydroxy hexadecanoid acid to 5-octanolide. Coriolic acid (13-hydroxy (Z9, E11)-octadecadienoic acid), the principle fatty acid of *Coriana nepalensis* seed oil, was a suitable precursor for the production of 5-decanolide by *Cladosporium suaveoleus*. Now that every major flavor company has developed its own process, more biochemical details of the formation of lactones have become known (Gatfield et al., 1993). The further oxidation of the intermediate 4-hydroxy decanoic acid from ricinoleic acid degradation yields 3,4-dihydroxy decanoic acid. This compound can cyclize to 3-hydroxy-4-decanolide. Dehydration gives rise to 2- and 3-decen-4-olide with mushroom and creamy/fruity odor attributes. The whole series of precursor hydroxy acids can be demonstrated in the *Candida* fermentation broth. Since 1985, lactone formation by *Fusarium poae* has been researched by a French group (Latrasse et al., 1994). Chirality and aroma properties of these fungal lactones were investigated using chiral GC and GC-O. The formation of 4-dodecanolide, (Z6)-4-

dodecenolide, and (E6)-4-dodecenolide were hypothetically related to actual fatty acid concentrations in this lipid accumulating species. Interestingly, the enantiomer ratio of lactones did not only depend on the type of nutrient medium, but also on the phase of growth. Large deviations of the concentrations of single lactones during the growth cycle were found. Obviously, more than one route to lactones is followed by this fungus.

Another fungal lactone with a long biotechnological history is 6-pentyl-2-pyrone (Table 5.2). Multistep chemosyntheses including reaction temperatures close to 500° C can afford this compound. In view of its strong, coconut odor and its involvement in biological pest control the biotechnological production has attracted the interest of several research groups. The compound inhibits the growth of phytopathogenic *Zygomycetes* and *Deuteromycetes*, and is accumulated in optimized bioprocesses in amounts up to 400 mg×L^{-1}. A C/N ration of 60 limited the growth of *Trichoderma harzianum* and favored lactone production (Serrano-Carreon et al., 1992). The strong stimulating activity of methyl ricinoleate again poses the problem of insufficient knowledge on biochemical regulation. Most recent work has confirmed the widespread potential of lactone formation in yeast and fungi: A fragrant 2-phenyl-5,6-dihydro-4-pyrone was isolated from surface cultured *Mycoleptodonoides aitchinsonii* at a level of 60 mg×L^{-1} (Akita Jujo Kasei, 1992). The compound with mushroom and cinnamon attributes was claimed for food and perfumes, and occurred de novo on a sawdust/bran medium. Various *Basidiomycetes* and *Saccharomyces cerevisiae* hydrogenated α, β-unsaturated lactones of *Massoi* bark oil (*Cryptocaria massoia*) in a patented process (van der Schaft et al., 1992b). High hydrogenation activity of the yeast was observed at 35 °C, pH 5.5, constant addition of glucose, and excess oxygen. A yield of 1.7 g of 5-decanolide was reported in an 8-hour/2-liter-scale process. Intact fatty acids were fed to strains of *Penicillium roquefortii* and transformed to some lactones (Chalier and Crouzet, 1992). The profile of volatiles was determined by the precursor profile and incubation conditions.

The biotechnology of volatile lactones had signalling character. Much of the success must be attributed to the low toxicity of the products on the biocatalytical systems. Thus, amounts of lactones beyond the 1 g×L^{-1} level can accumulate without significant inhibitory effect. A selection medium for overproducers may even contain a lactone as the sole carbon source (Nago et al., 1993).

With very few exceptions, the formation of volatile pyrazines has been restricted to bacteria. Concentrations in the g×L^{-1} range were accumulated by *Corynebacterium, Bacillus, Pseudomonas,* and *Streptomyces* species. The major product, tetramethylpyrazine, was observed in overproducers of acetoin, and, by analogy with the thermal route of generation, a condensation reaction with ammonia was postulated to afford enolamines as the immediate precursors. Mixed substituted pyrazines of *Pseudomonas* and *Streptomyces* were first detected as off-odors in contaminated foods and drinking water. These compounds can be used for positive flavoring applications in the appropriate chemical environment. The ring nitrogen is obviously derived from amino acids, and many microbial pyrazines still bear the alkyl side chains of valine, leucine and isoleucine in α-position. Recent work of Leete and coworkers (1992) summarized the occurrence of pyrazines in insects, microorganisms and food. Labeled amino acids and pyruvic acid were fed to *Ps. perolens* to rationalize the biogenesis of 3-isopropyl-2-methoxypyrazine. Amides of α-amino acids and glyoxal were excluded as precursors on the basis of NMR and MS-data. Though impressive biochemical and spectroscopic data were presented, there is still uncertainly on the origins of the two carbon unit required to react with the α-amino acid. The second nitrogen supposedly resulted from ammonia.

To stimulate the formation of tetramethylpyrazine, attempts were made to increase the concentrations of ammonia and acetoin in *Lactococcus lactis* (Kim and Lee, 1992). Other heterocycles, such as 3-methylfuran, or indol and skatol, were isolated from *Penicillium* and *Aspergillus* sp. (Boerjesson et al., 1992), and from *Coprinus picaceus* (Laatsch and Matthies, 1992). The nitrogen bicycles were also amenable by ruminal bacteria that contain active tryptophanases (Onodera et al., 1992).

Many of these heterocycles exhibit repellent odor characteristics as a pure compound. On dilution, they may exert desirable effects in various foods. Pyrazines deserve special attention, because they supply roasted, fried, or nutty odor notes, thus complementing the aroma of microwave- and other nonconventionally processed food.

Low molecular weight sulfur compounds belong to the most potent odorants with pleasant attributes near their detection threshold and repellent notes at increasing concentrations. As a result, the microbial generation of volatile thio compounds is often associated with the occurrence of off-odors. Methylated sulfur compounds were observed in anaerobic bacterial cultures fed with methoxylated aromatic substrates (Bak et al., 1992). The bacterial isolates were tentatively classified as *Pelobacter* and afforded methanthiol and dimethyl sulfide. Methylated volatiles failed to appear during growth on trihydroxybenzenes. The cyanobacterium *Microcystis aeruginosa* dominated the microflora of South African drinking water reservoirs and produced, besides geosmin (cf. Sect. 5.2), dimethyl trisulfide (Wnorowski and Scott, 1992). The same microorganism grew in inland waters of Japan and produced, among others, methyl isothiocyanate and isopropyl methylsulfide (Tsuchiya et al., 1992). A biotechnological approach will be hampered by the strict light requirements. Some species of *Penicillium* and *Aspergillus* were cultivated on oatmeal agar in cultivation vessels that allowed gas phase samples to collect (Boerjesson et al., 1993). GC-O revealed dimethyl disulfide as a major contributor to the musty off-odor of some of the species. In addition, 1-octen-3-ol, 2-methylisoborneol, geosmin, 1-methoxy-3-methylbenzene, and methylphenol were identified. The generation of the off-odor metabolites coincided with sporulation. Putrid spoilage of proteinaceous food is known to be caused by *Pseudomonas, Moraxella* and other species with high protease activity (Stutz et al., 1991). Volatiles, such as hydrogen sulfide and methyl oligosulfides have been suggested as indicators of contamination of food, although the smell would also tell us. In beef, the loss of primary volatiles as indicators of freshness appears more purposeful. In surimi, a kerosene-like off-odor was produced by either *Debaromyces han-*

senii and identified as styrene (Koide et al., 1992a), or by unidentified molds and found to be 1,3-pentandione from the degradation of added sorbate (Koide et al., 1992b). The unidentified mold was presumably a *Penicillium* species. Phenolic off-odors of beer and wine were identified as 4-vinylphenol and 4-vinyl guaiacol, generated from phenylpropanoic acids by the aid of yeast decarboxylases (Dugelay et al., 1993). During maturation of the wines the vinylphenols decreased in favor of an increase of the concentrations of ethoxy ethylphenols (cf. Sect. 2.3.2).

5.5 Degradation of Off-Odor

So called off-odor constituents regularly possess impact characters that can positively supplement, round off, or refine certain composed flavors. Examples are the products of the allylic oxidation of limonene and valencene, hydrogen sulfide (milk), methional (potato chip), and fatty aldehydes that are all off-odors in some, and essential compounds in other food flavors.

Morever, biocatalysts have indirectly improved the sensory quality of food by degrading off-odors to odorless or even to desirable products. A major problem in soybean processing is the lipoxygenase induced formation of n-hexanal and other volatiles that cause a 'beany' off-odor (Kobayashi et al., 1992; cf. Sect. 2.6). Several bacterial species effectively removed this off-odor by virtue of an aldehyde dehydrogenase activity. The fungus *Neurospora crassa* did not show this activity, but, nevertheless, improved the odor, as did *Alcaligenes faecalis*. A reductive mechanism was made likely by the beneficial influence of added NADH to crude enzyme preparations. A patent was granted on the use of *N. crassa* that claimed an excellent organoleptic quality of a soyabean protein cured in such a way (Ajinomoto, 1993).

Table 5.3. Recent applications of advanced biofilter systems

Biocatalyst(s)	Target	Reference
Not specified	ammonia, etc.	Colanbeen and Neukermans, 1992
Hyphomicrobium, Xanthomonas	H_2S, methane thiol, etc	Cho et al., 1992
Thiobacillus thioparus	thio compounds	Park et al., 1992
Not specified	methyl ethyl ketone	Deshusses and Hamer, 1993
Consortium, 8 sp.	methanol	Shareefdeen et al., 1993

An entire industry is making use of the flavor degrading properties of (usually mixed) bacterial cultures: Biofilters are successfully used on industrial scales in composting operations, in chemical plants including flavor manufactures, and in waste water facilities (Williams and Miller, 1992). Basically, the deodorizing system is a constant moisture solid state reactor, sometimes supplemented by gas scrubber, charcoal filters, etc. (Table 5.3). The solid phase conventionally consists of peat, straw, heather, or shredded wood. A bioscrubber with polyurethane foam as a packing material was described (Colanbeen and Neukermans, 1992). Oxygen transfer into the liquid film around the fixed bed particles and substrate load have been recognized as critical factors. A continuous supply of the microflora with metabolizable organic substrates is required for maintaining a constant biodegradative performance. The biofilter inoculum may be derived from the waste water treatment unit of the same plant.

5.6 Are Microbial Volatiles Secondary Metabolites?

Before bioflavors became an issue of increasing industrial interest, most researchers were satisfied with the structural identification of the chemical principles of an incidentally observed microbial odor. Later on, 'maximum' quantities of volatiles were determined at a more or less arbitrary point of the growth cycle, when an industrial application was envisaged. Only recently, more detailed kinetic data have been published that shed some light on fundamental conditions of the generation of microbial volatiles.

Growth and reproduction of microorganisms depend on a relatively small and uniform array of polyols, and carboxylic and amino acids. A network of universal biochemical pathways has evolved to interconvert the basic metabolites into other essential ones, and to assemble them into larger structures. Both anabolic and the energy generating catabolic pathways follow common chemical patterns, such as activation, condensation, hydrolysis, isomerization, or redox reactions. Metabolites involved into these essential biochemical functions are defined as primary. This definition implies that metabolites accumulating in batch cultures after the phase of active growth (cf. Sect. 4.1) are lacking this primary metabolic role and have, thus, to be termed secondary. In fact, many microbial volatiles occur after all growth has closed. Primary metabolites start getting accumulated and are, then, funneled into secondary pathways. For example, the de novo synthesis of styrene by *Penicillium caseicolum* started only after glucose was completely exhausted (Spinnler et al., 1992) However, idiophasic production is not necessarily related to the above definition, and volatiles without known metabolic function can as well be synthesized in the trophophase. A perfect example for this reversed situation were submerged cultured cells of *Mycena pura* that shown production of citronellol from the first day of cultivation to the middle of the linear phase (Abraham et al., 1993). Citronellol concentration dropped rapidly and reached lag phase levels when the cells entered the stationary phase (Fig. 5.13). As the

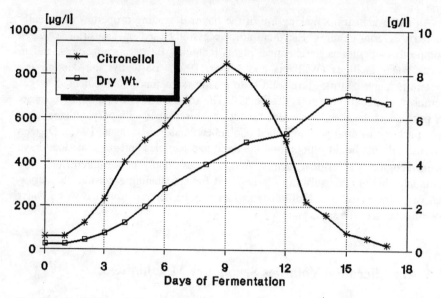

Fig. 5.13. Formation of citronellol in batch culture of *Mycena pura* (Abraham et al., 1993)

volatile monoterpenol was continuously stripped off during the bioprocess, a metabolic function could be largely excluded. A massive loss of carbon is not compatible with cellular economy anyway. In the same culture, acetoin, benzaldehyde, *p*-hydroxybenzaldehyde, and 2-phenyl ethanol culminated at the end of the linear phase. Some volatile phenols and benzoate esters peaked during the idiophase. Similarly, *Hansenula* yeasts produced the majority of total volatiles, namely the fusel oil constituents during active growth, and ester formation occurred in a subsequent, timely differentiated phase of carbon depletion (Janssens et al., 1988). De novo synthesis of lactones by *Fusarium poae* ran parallel with the formation of biomass (Latrasse et al., 1994).

Eight carbon alcohols and carbonyls frequently occur in stationary phase cultures of fungi (Spinnler et al., 1992; Abraham et al., 1993). This phenomenon should not be discussed in terms of primary vs secondary metabolism. Lysis of cells during senescence commences in the late stationary phase and brings about the contact of fungal lipoxygenase with its fatty acid substrates. During earlier phases of the growth cycle mechanical or chemical damage of the cells will start the same reaction sequence. Thus, the abundance of 1-octen-3-ol is regarded as being indicative of the vitality status of fungal cells.

Metabolic imbalances are created by the variants of precursor feeding. The microbial cell, in an attempt to bring back overflowing metabolic pools to homeostatic equilibria, activates conjugating and volatilizing mechanisms to get rid of cytotoxic intermediary concentrations. Examples were the spontaneous accumulation and excretion of benzyldehyde in phenylalanine supplemented cultures of *Ischnoderma benzoinum* (Krings, 1994), or the castor oil based bioprocesses for lactone formation (Gatfield et al., 1993). Some transformations/conversions (cf. Sect. 6) worked better with stationary phase cells; however, little is known on the inducability of the respective pathways in other phases of cellular development.

The uneven distribution of the potential to form volatiles in the various microbial taxa is obvious. Organisms such as the enterobacteria adapt rapidly to changing environments and grown well in the absence of chemicals inhibitory to other species. Soil streptomycetes have to be equipped with a flexible chemical defense system to survive in a heavily colonized environment (cf. Sect. 1.5). Acid forming bacteria control the competing flora by the harsh pH conditions and, more and more recognized, by excreting bactericidal proteins (cf. Sect. 2.5). Filamentous fungi produce a wealth of volatiles. Presumably their function is to attract insects as spore-carriers, to repel predators, or generally to mediate allelopathic activity. Unicellular fungi exhibit a lower propensity for generating volatile compounds and rely, under conditions of limited oxygen, on the cytotoxic activity of higher levels of ethanol.

Restricting the discussion to de novo synthesized compounds the respective definition of terms decides into which metabolic category a microbial volatile is to be placed. As many microorganisms obviously exist without producing volatiles at all, one might completely negate any primary metabolic functions. Consequently, microbial volatiles would be secondary metabolites. However, it has become clear that the enormous synthetic potential of the filamentous fungi and some other genera was overlooked for such a long time because of the inappropriate experimental conditions of growth. Many researchers now start optimizing proliferation and, then, subsequently establish product formation concurrent to the apparent trophophase on a much higher level of bioactive cell mass. This would distinguish volatile compounds from classical idiolites, such as antibiotics. Finally, considering the essential role of volatile chemical messengers in spore distribution, defense, or in other vital ecological functions (Tables 1.2, 1.3, 1.5), certain volatiles definitely occupy a primary functional position. Artificially constructing such an ecological motive, for example by co-cultivating elicitors, can stimulate the production of bioflavors. These considerations cross the border from conceptual mind games to real world bioprocess optimization.

6 Biotransformation/Bioconversion

Many volatile target compounds are not amenable to a microbial de novo synthesis in anything like acceptable yields. Both constitutive or inducable microbial enzymes, however, turn over biotic intermediates and even xenobiotics in single step (biotransformations) or multistep reactions (bioconversions) to products more valuable from a flavor point of view. Biocatalysts can enlarge, degrade, or modify the substrate, thereby supplementing or replacing chemosynthesis. Particularly attractive is the breakdown of complex natural products to volatiles. Recent oil tanker accidents near the Shetlands and in Prince William Sound, Alaska, have unintentionally demonstrated to a broader public the impressive biodegradative capabilities of the indigenous microflora (Venosa et al., 1992).

As outlined above (cf. Sect. 3.3), specificities and physiological conditions of biochemical reactions favor their application in the field of volatile flavors. Biocatalysis competes best with chemical catalysis in the following types of reactions:

- Functionalization of chemically inert carbons. Examples were regioselective hydroxylations of terpenoids.
- Selective modification of one functional group in multifunctional molecules. Hydrogenation of one out of several double bonds was described above (Fig. 3.2).
- Introduction of chirality. Transformation of prochiral carbons, for example to hydroxy carboxylic acids or volatile 2-alkanols, or the lipoxygenase induced hydroperoxidation of fatty acids could be cited.
- Resolution of racemates. This is industrially used to separate racemic α-amino acids and menthol acylates.

The beginning of modern transformation biotechnology is usually dated back to Neuberg's classical work with yeast-catalyzed reactions of terpenes and benzaldehyde. The acyloin type condensation of benzaldehyde and acetaldehyde is still used to produce an intermediate of ephedrin. After the first bioactive steroids were structurally characterized, microbial dehydrogenation, side chain degradation, and hydroxylation of steroid skeletons led to a cost-effective production of building blocks for subsequent chemoderivatization to pharmaceuticals (Vischer and Wehstein, 1953). These findings had a strong impetus on further research. In 1960, the group of Bhattacharyya launched a series of classical publications on the microbial transformation of volatile terpenoids; *Aspergillus niger* and, later on, *Pseudomonas* strains were the catalysts of choice (Dhavalikar and Bhattacharyya, 1966). This work used a variety of mono- and sesquiterpene hydrocarbons, sometimes as the sole source of carbon, to evidence a degradation to carbon dioxide and water that was accompanied by the occurrence of partially oxidized, volatile compounds. The following reaction steps were observed:

- Oxidation of methyl groups, and of primary, secondary, tertiary (3-carene), and phenolic (thymol) hydroxy groups
- Hydroxylation of allylic carbons, sp^2-carbon (to yield oxo compounds), dihydroxylation of vicinal positions (limonene)
- (De)hydrogenation, epoxidation, isomerization, and hydration of carbon double bonds
- Cleavage of carbon/carbon and carbon/oxygen bonds

In sharp contrast to the current use of biocatalysts in steroid transformation, the possibility of applying the same technique to lower terpenoids has been largely neglected by industry. Difficulties encountered are the incompatibility of lipophilic substrates and aqueous bioprocess media, the generation of useless or difficult to separate by-products, the lack of predictability if new strains are used, and the sufficient availability of many volatiles terpenes from plant sources. Novel concepts to overcome some of the inherent drawbacks will be discussed later (cf. Sect. 11.1.1)

A survey over the present literature shows that filamentous fungi and prokaryotes share a preference for converting volatile terpenes. The absolute number of genera specialised in this metabolism is surprisingly small (Table 6.1). Yeasts, with few exceptions, are not applicable to terpene bioconversions (cf. Sect. 2.3.1). To improve the chances of success, an organism should be selected that is known to perform the desired type of reaction on other compounds. If a mixed culture isolated from a natural habitat is used instead of a commercial strain, the member with the highest activity is easily selected by enrichment media containing the substrate as a sole carbon source. Eukaryotic organisms are usually more sensitive to terpenes and must be adapted stepwise. The surroundings of plant terpene producers should be considered as a particular ecological niche for adapted strains.

Much of the recent work on microbial transformations of volatile terpenes was conducted by the group of Kieslich and Abraham (Kieslich et al., 1985). A database covering worldwide information on microbial transformations is currently being set up (Kieslich, pers. comm.). Hence, discussion can be limited to representative examples.

Mycobacterium smegmatis was most suitable among more than 100 strains tested to convert the activated methyl groups of (+)-2- and (+)-3-carene to chaminic acids, insect repellents difficult to obtain by chemosynthesis (Stumpf et al., 1990). Carenones and a compound with a cleaved cyclopropane ring were identified as volatile by-products (Fig. 6.1). Geranial, neral, and further alkadienals reacted to hydroxyketones in fermentations of *Mucor circinelloides* (Stumpf and Kieslich, 1991). The acyloins were immediately reduced to (2S,3R)-diols. Reduction of C=C, and C=O bonds, cyclizations, and further catabolic reactions yielded several other products. Previously, acyloin formation was believed to be restricted to yeasts. The spectrum of transformable aldehydes was greater in the *Mucor* culture

Table 6.1. Preferred microorganisms for the bioconversion of volatile terpenes

Prokaryotes
Bacillus   hydroxylations, diols, resolution of racemates
Corynebacterium   hydroxylations
Mycobacterium   oxidations, resolution of racemates
Nocardia   hydroxylations, resolution of racemates
Pseudomonas   all kinds of oxidative degradation
Streptomyces   hydroxylations, epoxidation, cyclisation

Yeasts
Candida, Hansenula   ⎫   resolution of racemates, stereo-
Rhodutorula   ⎬   selective reductions, formation
Saccharomyces   ⎭   of C-C bonds

Higher Fungi
Absidia   resolution of racemates, hydroxylations
Alginomonas   resolution of racemates, hydroxylations
Aspergillus   all kinds of oxidative degradation
Botrytis cinerea   hydroxylations, hydrogenations
Chaetomium   oxidative pathways
Cladosporium   diols
(Lasio) Diplodia   hydroxylation, epoxidation, diols
Fusarium   hydratation
Gibberella   hydroxylation, resolution of racemates
Mucor   epoxidation, redox reactions
Penicillium   redox reactions, hydroxylations
Trichoderma   resolution of racemates

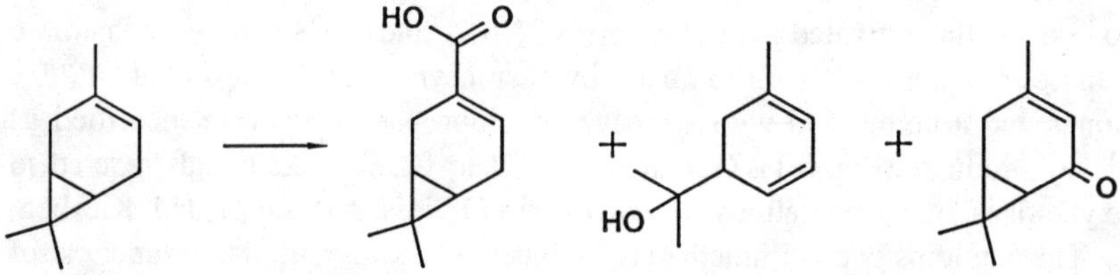

Fig. 6.1. Transformation of (+)-3-Carene by *Mycobacterium smegmatis* (Stumpf et al., 1990)

than in comparable incubation studies using baker's yeast. ω-hydroxylated terpenoids constitute another group of volatiles that are important flavor chemicals or intermediates not easily accessible by chemical means (Arfmann et al., 1988). Nerolidol and related structures varying in configuration and molecular weight were used as substrates. *Aspergillus niger* or *Rhodococcus rubrapertinctus* gave at least one ω-hydroxylated product of each of the substrates. An extended screening of microorganisms was required indicating that thermodynamically unfavored reactions pose problems in living systems as well as in a chemically catalyzed process. Fungal strains were preferred for the transformation of mono-, bi-, and tricyclic sesquiterpenes (Abraham and Stumpf, 1987; Abraham et al., 1989). Humulene was transformed by *Chaetomium cochlioides* and *Diplodia gossypina* to afford over thirty new humulanes. Caryophyllene, calarene and globulol transformation was also studied. Whenever possible, fungi were used because of a less rapid catabolism of the oxygenated volatiles formed. As a result, many transformation products accumulated in reasonable yields. Less volatile, vicinal diols were often reported and appear to be regular intermediates in a general microbial degradation pathway (Abraham et al., 1992; Fig. 6.2).

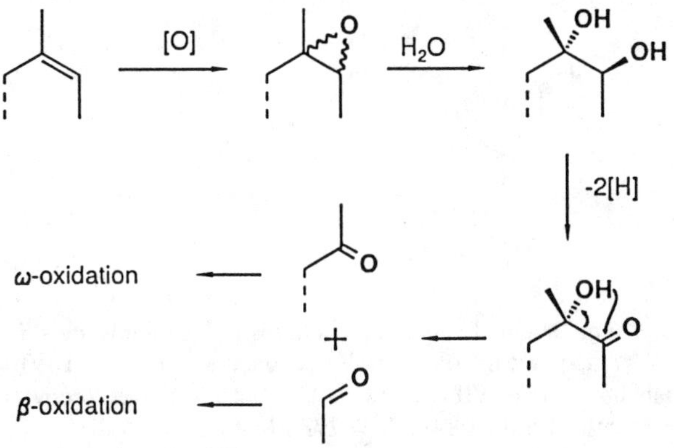

Fig. 6.2. Oxidative degradation of an isoprene unit via 1*S*,2*S* (or 1*R*,2*R*)-diols by microorganisms and in vitro plant cells

6.1 Monoterpenes

Monoterpene hydrocarbons dominate in most essential oils. They are usually separated from the oils by rectification because of low odor activity, high hydrophobicity, and high tendency to autoxidize and polymerize. This renders abundant monoterpenes, such as α-pinene and limonene, an industrial waste and an inexpensive starting material for chemical and biochemical transformations.

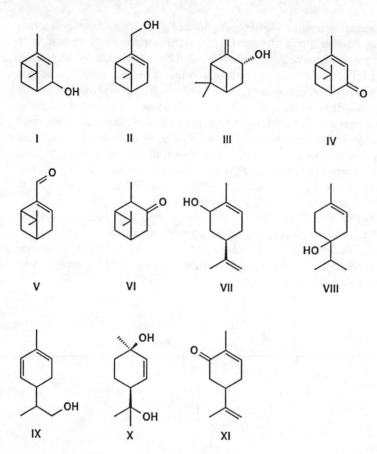

Fig. 6.3. Volatiles in α-pinene supplemented submerged cultures of basidiomycetes. I = Verbenol, II = Myrtenol, III = Trans-pinocarveol, IV = Verbenone, V = Myrtenal, VI = 2,6,6-Trimethylbicyclo (3.1.1)heptan-3-one, VII = *trans/cis*-Carveols, VIII = Isopropyl-4-methylcyclohex-3-enol, IX = Mentha-1,5-dien-8-ol, X = *E/Z*-Menth-2-en-1,8-diol, XI = Carvone

Biotransformations of α-pinene (Fig. 6.3) were generally characterized by:

- oxidative attack of the allylic position,
- oxidative cleavage of the cyclobutane ring, and
- oxidation of terminal carbons in 7/10-positions.

High substrate volatility and toxicity prevented high yields when *Aspergillus niger* or *Pseudomonas* sp. were used. Parallel autoxidation demanded an efficient biocatalyst to reduce the time of incubation. On the other hand, the fast prokaryotes usually yielded a lot of diols and monoterpene acids with little or no flavor value. Selectivity and yield were improved using *Acetobacter methanolicus* under conditions of limited growth and in the presence of a second carbon source (Repp et al., 1990). Except from the monoterpene acids and some of the diols, almost all of the microbial catabolites of α-pinene were recovered from basidiomycete cultivations (Fig. 6.3). Allylic oxidation was regularly and immediately found. Volatiles resulting from an attack of the double bond were identified in smaller concentrations. The fission of the carbon bridge largely depended on the basidiomycete species used.

While *Pseudomonas* sp. transformed β-pinene in a similar way, *A. niger* acted more specifically on the molecule and yielded mainly pinocarveol, pinocarvone and myrtenol. The same preference was found with basidiomycetes. However, in spite of the theoretically identical intermediary carbocations of α- and β-pinene the spectra of products differed significantly. It could be concluded that the relative stability of the isomer *p*-menthadiene cations after opening of the carbon bridge determined the structure and amounts of the resulting monoterpenols (Fig. 6.4). Such 1,2- and 1,3-hydride shifts and Wagner-Meerwein rearrangements are not uncommon in the biosynthesis of monoterpenes. The unique potential of basidiomycetes is underscored by the identification of the cineols and of fenchol, impact volatiles that have not been described as microbial products of pinenes before (Fig. 6.5). In view of the widespread occurrence of cytochrome P-450 monooxygenases and the popularity of essential oil constituents in treating the common cold, rheumatism, etc., microbial transformations have also been looked at as models to study human terpene metabolism. Feeding studies, for example on rabbits (Ishida et al., 1991), have demonstrated allylic oxidation and other similarities, but major dissimilarities as well. The different microbial responses may be paralleled by different responses of mammals, and all toxicological conclusions should be made with great care.

Limonene is probably the best studied monoterpene in biotransformation research. Four major pathways have been distinguished:

- allylic oxidation to *cis*- and *trans*-carveols,
- oxidation of the methyl substituent to the perilla compounds
- reductive epoxidation of the Δ 8,9-double bond to α-terpineol, and
- epoxidation of the ring double bond to the diol (Fig. 6.2).

Many of the intermediates branch off to further volatiles, some of which possess industrial impact (carvone, dihydrocarvone, perillaldehyde, α-terpineol). To reduce the recurrent problem of low solubility and physiological compatibility of limonene, attempts to increase the yields of products included the use of *Penicillium* isolated from citrus peel, or the use of bacteria able to utilize alkanes as a carbon source.

Due to the manifold studies, biotransformation of limonene can serve as an example for a more detailed stereochemical discussion. An incubation of *R/S*-limonene mixtures with *Penicillium digitatum* exceeded expectations and afforded optically pure (+)-(4R)-α-terpineol in good yields (Kieslich et al., 1985). This selective hydration was explained by the adsorption of the (S)-substrate into the mycelium or, more likely, by a rapid and selective degradation of the other enantiomer. On the other hand, *Chaetomium* sp. transformed *S*-limonene into a mixture of (6S)- and (6R)-carveol. A planar, intermediary carbocation must be assumed

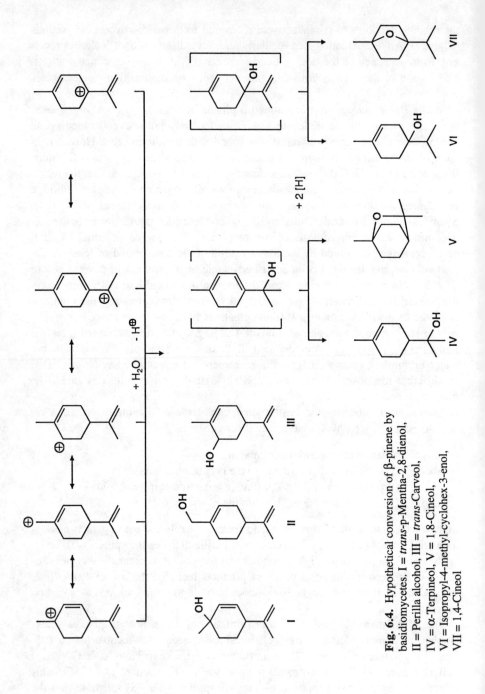

Fig. 6.4. Hypothetical conversion of β-pinene by
basidiomycetes. I = *trans*-p-Mentha-2,8-dienol,
II = Perilla alcohol, III = *trans*-Carveol,
IV = α-Terpineol, V = 1,8-Cineol,
VI = Isopropyl-4-methyl-cyclohex-3-enol,
VII = 1,4-Cineol

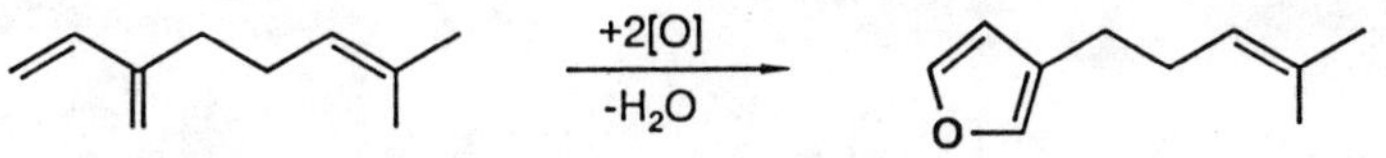

Fig. 6.5. Fungal conversion of myrcene to perillene, one of the sensory principles of rose oil (Busmann, 1994)

that was equally well hydroxylated on both sides of the ring. Racemates and pure limonenes were converted according to expectations by *Aspergillus cellulosae* (Noma et al., 1992). For example, *R*-limonene was converted to *R*-configurated isopiperitenone, *cis*-carveol, and perillalcohol. The opposite was found in basidiomycete cultures. Upon incubation with pure (*R*)- or (*S*)-limonenes, all of the four stereoisomers of carveol were separated on a chiral capillary column (Table 6.2). It was believed that the para-positioned ring carbon formed a sp^2-hybridized, trienoic carbocation that could, by analogy to the derivatization of myrcene, cyclize to the racemate (Fig. 6.6).

Myrcene with its 1,3-diene system is prone to rapid polymerization and has not been well investigated in biotransformations. Basidiomycetes produced acyclic volatiles, such as myrcenol or nerol, and monocycles, such as perillalcohol, dihydrocarvone and menthandiols. Myrcenol dominated because of the obvious stability of the intermediary (tertiary) carbocation. The monoterpenols can formally be derived from the above (Fig. 6.6) (primary) carbocation at the terminal carbon that is not stable enough for hydroxylation and cyclizes instead in a Michael type reaction (Fig. 6.7). Such a formation of new carbon-carbon bonds is a rather rare event in microbial transformations. The hypothetic carbocation, however, is sup-

Table 6.2. Distribution of isomer carveols[a] upon incubation of pure limonenes with basidiomycetes (Busmann, 1994)

Species/Substrate	(+)-cis-carveol	(-)-cis	(-)-trans	(+)-trans
(+)-limonene:				
Ganoderma applanatum	35.4	18.4	25.8	20.3
Marasmius alliaceus	22.3	29.0	30.0	18.6
Pleurotus sapidus	10.5	23.2	36.4	29.9
(-)-limonene:				
G. applanatum	11.7	17.3	33.6	38.6
M. alliaceus	15.3	21.8	27.3	33.5

[a]As trifluoroacetates on octakis (6-*O*-methyl-2,3-di-*O*-pentyl) – γ-cyclodextrin

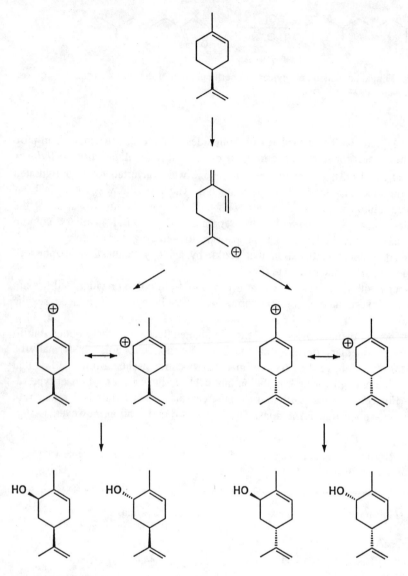

Fig. 6.6. Hypothetical formation of all carveol isomers from pure (-)-*S*-limonene (Busmann, 1994)

ported by the occurrence of camphor and lavenderlactone, and of menthol during incubations of linalool the presence of *Pseudomonas*, and of citronellal the presence of *Pseudomonas* or *Penicillium*. Both substrates and the myrcene carbocation have similar steric and electronic features in common. Recent work focussed on conversion of oxygenated monoterpenes, such as 1,4-cineol, perillalcohol, and geranylacetone (Rosazza et al., 1988; Cardillo et al., 1992; Madyast-

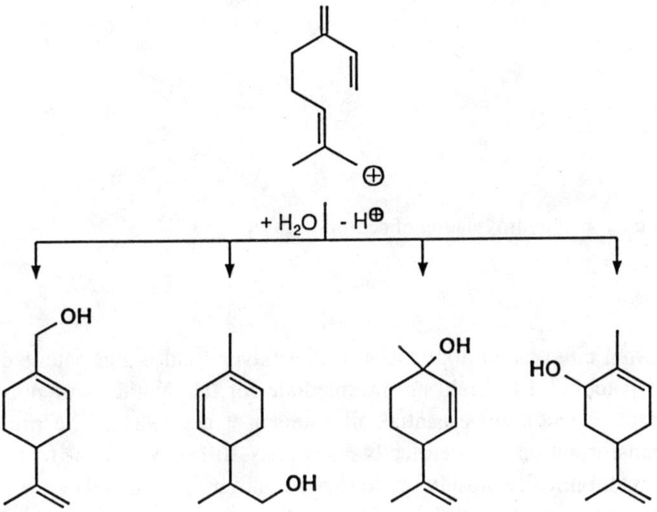

Fig. 6.7. Monocycles from the conversion of myrcene by basidiomycetes (Busmann, 1994)

ha and Gururaja, 1993). The stereochemical aspects were emphasized and deuterated substrates and chiral separations were applied to elucidate mechanistic details. As a result, stereoselective hydroxylations and regioselective hydrogenations were frequently observed in these bioprocesses. However, biocatalyst, like all catalysts, must obey a number of fundamental stereochemical and thermodynamic rules. This explains the frequent overlap of preferred products in biological and chemical systems.

6.2 Sesquiterpenes

Sesquiterpenes are of considerable value in flavor and pharmaceutical applications. A recent review has shown that non-functionalized sesquiterpene hydrocarbons with their abundant equivalent substructures often yield highly complex mixtures of bioconversion products (Lamare and Furstoss, 1990; Abraham et al., 1989). Monooxygenated sesquiterpenes are more water soluble and provide an anchoring group that obviously helps reducing the number of suitable biocatalysts and specificities. However, exceptions to this rule can be found easily: α-Cedrene was regiospecifically hydroxylated to produce (R)-10-hydroxycedrene by a *Rhodococcus* sp. (Takigawa et al., 1993), and regular (8-hydroxy)-cedrol was transformed to 2S-hydroxycedrol as the only metabolite by *Bacillus cereus* (Maatooq et al., 1993). *Streptomyces griseus*, by contrast, yielded numerous hydroxy-, oxo-, and dihydroxycedrols. The majority of bacteria and fungi examined preferred the 2- and 12-positions for the introduction of the hydroxy function (Fig. 6.8; Hanson

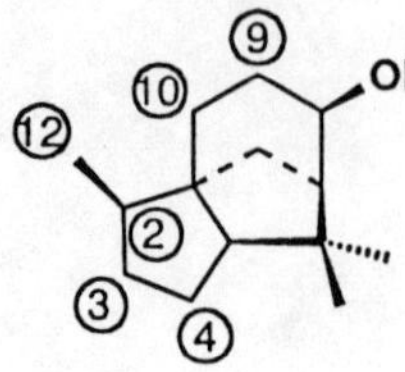

Fig. 6.8. Positions of microbial hydroxylation of cedrol

and Nasir, 1993). Often cited as an application in the flavor field is the patented 10-hydroxylation of patchoulol to yield an intermediate for the synthesis of nor-patchoulol, the impact of patchouli essential oil (Suhara et al., 1981). The pre-dictability of the transformation of chemically inert sites still leaves a lot to be desired, even when taxonomically closely related species are used (Fig. 6.9).

The microbial degradation of norisoprenoids, such as for the preparation of tobacco flavorings has received attention by several groups (Mikami et al., 1981; Krasnobajew and Helminger, 1982). In a patented process end of log phase cells of *(Lasio) Diplodiatheobromae* transformed β-ionone in a Baeyer-Villiger type oxidation to the main product β-cyclohomogeraniol and other, previously un-known metabolites (Fig. 6.10). Hydroxylation of the non-activated 2-, 3- and 4-positions was also observed, particularly in the presence of *Aspergillus niger*. A series of related substrates including β-damascone was also submitted to this process and found to react along similar routes to odor active trimethylcyclohexa-ne derivatives.

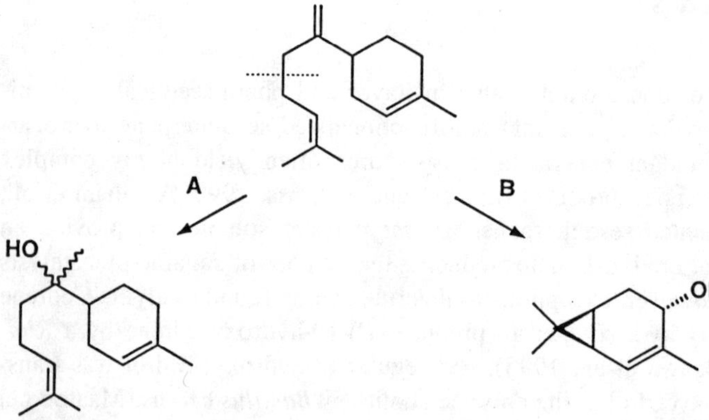

Fig. 6.9. Biotransformation of β-bisabolene by A) *Marasmius alliaceus*, and B) *Mycena pura* (Busmann, 1994)

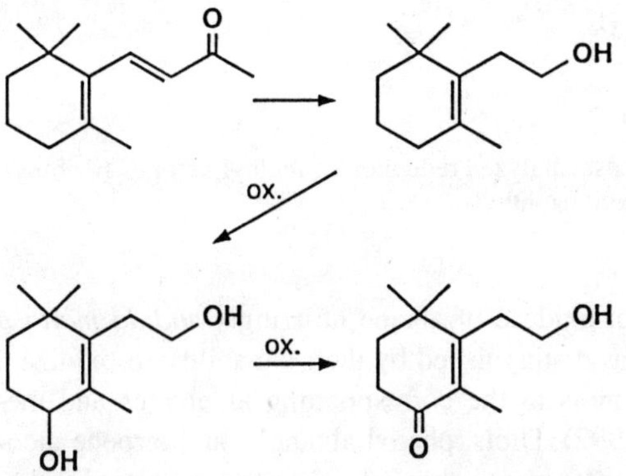

Fig. 6.10. Bioconversion of β-ionone yielded tobacco flavorings (Krasnobajew and Helminger, 1982)

6.3 Fatty Acid Derivatives

A lot of experience has been accumulated on the use of *Saccharomyces cerevisiae* as a catalyst and reagent for the enantioselective hydrogenation of the carbon/carbon double bond of α, β-unsaturated, methylated carbonyl compounds, and for the stereoselective reduction of carbon/oxygen double bonds to yield chiral carbinols. As cultures of yeast cells are easy to obtain and maintain, these selective steps have been widely applied to produce pure synthons on the route to pheromones, vitamines and other bioactive products. The reduction of keto esters, keto acids, and generally carbonyls has aroma implications. The hydrogen transfer to a prochiral keto group is mainly sterically controlled, as expressed by Prelog´s rule: Transfer of the hydrogen to the Re-face results in the formation of an (S)-alcohol (Fig. 6.11). Studies on the reduction of 3-keto esters have suggested the presence of two or more competing enzymes of opposite chiral preference. Dependent on substrate concentration, immobilization of the yeast, and the incubation conditions in general, yield and enantiomeric excess varied.

As discussed in the preceding chapter, volatile lactones can be synthesized de novo, via bioconversions, or via direct cyclization of precursor hydroxy acids. *Nocardia cholesterolicum* provided access to the 10-hydroxy acids of oleic, linoleic, and linolenic acid (Koritala and Bagby, 1992). Reaction conditions were optimized to give conversion yields of about 80%. Other strains of *Nocardia* performed similarly well in this hydration reaction, while a *Staphylococcus* mainly produced the 10-keto acid (Lanser, 1993). By contrast, *Bacillus pumilus* converted oleic acid to a mixture of 15-, 16-, and 17-hydroxy-9-octadecanoic acid (Lanser et al., 1992). Some of the hydroxy acids may be used as substrates in lactone yielding bioprocesses, thereby circumventing the slow lipoxygenase step otherwise required.

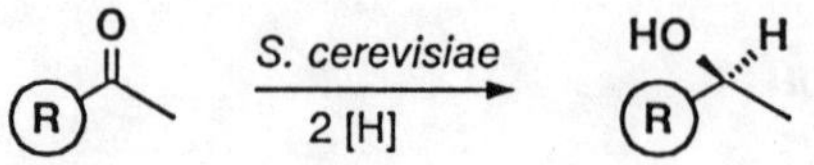

Fig. 6.11. Stereochemical course of yeast catalyzed reduction of methyl ketones (R=butyl, 1-pentinyl, 3-hexenyl, phenyl, cyclohexyl, naphthyl)

A membrane-bound fatty alcohol oxidase of alkane utilizing *Candida maltosa* and other alkane utilizing yeasts was distinguished by their capability to oxidize a broad spectrum of 1-and (2*R*)-alkanols to the corresponding aldehydes and methylketones (Mauersberger et al., 1992). Diols, phenyl alkanols, and terpene alcohols were accepted substrates, but the enzyme showed decreased activity. Further oxidation of the aldehyde to the carboxylic acid was excluded on the basis of stoichiometric oxygen consumption. In a patented process, microorganisms were first cultivated anaerobically to convert sugar, followed by an aerobical phase in the presence of a short-chain precursor alkanol (Phillips Petroleum, 1991). The carboxylic acids, after drying of the entire biomass, imparted buttery and cheese odors. Oxidative and reductive biotransformations of cycloalkanones were obtained by using *Acinetobacter calcoaceticus* (Sandey and Willetts, 1992). Cofactor regeneration is necessary to operate any of these redox processes economically. For example, the cofactor for the stereoselective reduction of 3-keto esters by *Geotrichum candidum* was NADPH (Patel et al., 1992). This cofactor was recycled by glucose dehydrogenase in a second, complementary enzymatic reaction.

6.4 Aromatics and Heterocycles

The transformation of benzyl alcohol by *Pseudomonas cepacia* (Horiuchi et al., 1993), and the *p*-hydroxylation of methyl cinnamate by the brown-rot fungus *Lentinus lepideus* (Ohta and Shimada, 1991) may serve as further examples that soil bacteria and higher fungi are the organisms of choice for the generation of volatile aromatics. The carboxylic function of several benzoic, cinnamic, and phenyl acetic acid derivatives was most effectively reduced, with almost no side products, by higher fungi (Arfmann and Abraham, 1993). Within two to three days up to 80% of the substrate was converted. Dicarboxylic acids were not reduced. The reduction of acetophenone to 1-phenyl ethanol by *Aspergillus niger*, isolated from dried bonito (*Katsuobushi*), was asymmetric (Doi et al., 1992). Seven of the strains tested produced the *R*-isomer, while one strain gave predominantly the *S*-isomer. These results brings discussion back to imponderabilities and opportunities of biocatalysts, and to the importance of the genuine microflora of traditional foods. The positional isomer 2-phenyl ethanol was the substrate of acetic acid bacteria to produce the valuable 2-phenyl acetaldehyde (Manzoni et al., 1993). Different parameters of the bioprocess were optimized to achieve yields > 50%.

Furfural and 5-hydroxymethylfurfural are indicative of a thermal treatment of sugar containing foods and pose problems in the microbial degradation of sugar cane molasses, etc. *Saccharomyces cerevisiae* under low aeration (Villa et al., 1992; Diaz et al., 1992), and enteric bacteria under both aerobic or anaerobic conditions (Boopathy et al., 1993) converted furfural to furfurol in good yields. The thio analog is an impact component in coffee and roast beef aroma. It was prepared by coupling cystein to furfural and microbial cleavage of the cystein-furfural conjugate into furfurythiol, pyruvic acid, and ammonia (van der Schaft et al., 1994; Fig. 6.12). *Enterobacter cloacae* was selected as one of many bacteria strains that showed carbon-sulfur-lyase activity. Upon investigating the key parameters the optimized process was patented and claimed to yield up to 50% conversion of substrate. Continuous removal of the volatile was again found essential due to product inhibition.

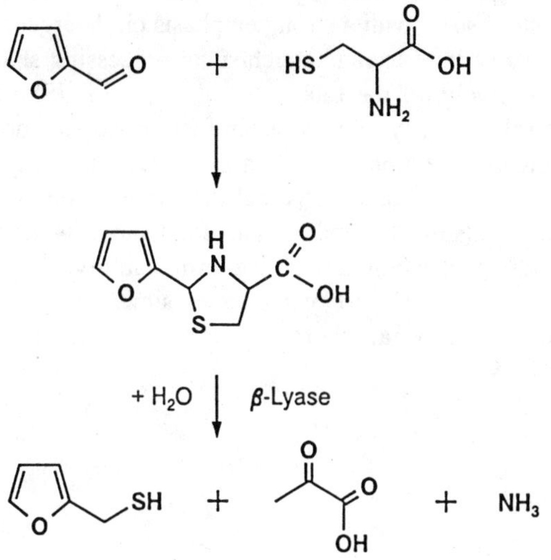

Fig. 6.12. Biosynthesis of furfurylthiol by *Enterobacter* (van der Schaft et al., 1994)

7 Enzyme Technology

In the past decade, biocatalyzed reactions have almost developed into a fashion in synthetic organic chemistry (Faber, 1992). If a given organic molecule is to be modified in a single-step, cofactor-independent reaction, no plausible reason remains for supporting all the other biomass and metabolic energy delivering pathways of an intact cell (cf. Table 3.1). Of the estimated 25 000 enzymes present in nature, about 2800 have been classified, and about 400, mainly hydrolases, transferases, and oxidoreductases have been commercialized. Less than 50 different enzymes are used on a larger industrial scale with a strong emphasis on detergents and food processing. In 1992, the dairy industry and starch/sugar-processing shared 56 mio and 44 mio US$, respectively, of the total sales of 350 mio US$ of technical enzymes in Europe. The vast majority of enzymes in food processing are hydrolases, such as amylases, proteases, pectinases, cellulases, pentosanases, invertase, and lactase. Immobilization techniques, such as inclusion in gels, entrapment in microcapsules, or covalent or adsorptive binding onto solid supports have improved technical aspects of handling, recycling, and long-term stability. Since microbial enzymes have become an integrated part of processes aimed at value-added food, it would seem obvious to propagate their use for the generation of volatiles (Whitaker, 1992).

7.1 Lipases

The ubiquitous lipid hydrolases (EC 3.1.1.3) bear flavor aspects in liberating volatile fatty acids from the corresponding triacylglycerols and further esters, and in accepting different acyl and alkyl moities in reversed hydrolytic reactions. In the first case, a perfectly water soluble biocatalyst acts on lipophilic substrates that selforganize in water as monomolecular films, bilayers, micelles, liposomes, or emulsions. In the second case, hydrophobic substrates are reacted at the (hydrophobic?) active site of a hydrated enzyme protein in a microaqueous environment. This heterogeneous phase type of catalysis ('interfacial enzymology') has been explained on the basis of X-ray crystallographic studies. A converging evolution at the molecular level has generated microbial and human enzymes with very similar catalytic sites (asp-his-ser) that are, in the homogeneous phase, buried by a small protecting helix. A conformational change as induced by a lipophilic interface ist thought to open the peptide lid, thus giving way to the active triad.

This lid causes the enzyme's inactivity in aqueous phase, the activatory and inhibitory effects of bile salts, fatty acids, proteins and other bipolar compounds, and, except from severe treatments such as photooxidation, a superior resistance to chemical inhibitors.

Purified lipases often show complex patterns after isoelectric focussing, but this heterogeneity is obviously due to varying degrees and positions of glycosilation of the core protein. Pre-pro-lipase and pro-peptides are now studied in detail by genetic engineering methods (cf. Sect. 8.2.1). Lipases with very different specificity and stability have been reported. For example, fungal lipases were active at pH 10 (Stöcklein et al., 1993), at pH 2 (Hädrich-Meyer and Berger, 1994), and in organic solvents (Shimada et al., 1993). Regarding the broad range of properties it appears impossible to compare lipases of different origins in a standardized activity assay. However, the release of free fatty acids from olive oil at 37 °C and pH 7.8, and the release of butanoic acid from glycerol tributanoate at 30 °C and pH 6.8 have been agreed upon as analytical parameters to distinguish lipolytic from esterolytic activity (Table 7.1). Both lipase and esterase activity of well-known producer species differ by orders of magnitude. Highly active lipases usually possess good esterase activity. Esterase activity on a unit per mass basis is often higher than lipase. Very low activity data were reported for the *Penicillium roquefortii* lipase, matching previous experience that lipolytically active dairy strains jeopardize the sensory balance of the product (cf. Sect. 2.5).

Table 7.1. Lipolytic and esterolytic activity of microbial lipases (data taken from commercial sources)

Species	Lipase [U×g^{-1}]	Esterase [U×g^{-1}]
Chromobacterium viscosum	144 000	2 650 000
Candida cylindracea	85 700	53 600
Rhizopus javanicus	36 000	39 400
Candida lipolytica	10 500	150
Rhizopus japonicus	8 750	13 300
Pseudomona fluoresceus	8 500	26 300
Geotrichum candidum	4 650	170
Mucor miehei	4 200	22 000
Aspergillus niger	1 580	940
Penicillium roqueforti	290	625

7.1.1 Lipolysis

Milk fat was effectively hydrolyzed by *Candida cylindracea* or *Rhizopus arrhizus* lipase to obtain butter and cream notes (Chen and Yang, 1992). Immobilization in photo-crosslinkable resin increased the release of the aroma principles, butanoic and hexanoic acid. A lipase of *Pseudomonas aeruginosa* was more active on short-chain vs. long-chain triacylglycerols (Chartrain et al., 1993). The lipase was successfully used to prepare an ester acid from a diester. Sonnet (1993) reported the lipase catalyzed resolution of 2-methyloctanoic acid via isopropylidene glycerol and glycidol esters. The effect of alcohol configuration on the stereobias of the hydrolysis was modest and consistent with the proposed catalytic mechanism. This principle of kinetic resolution of racemic precursors to synthesize volatile alkanols and acids has found broad application in flavor chemistry.

7.1.2 Kinetic Resolution of Racemates

L-menthol, a chiral aroma molecule with impact character, is used for flavouring candies, tobacco, chewing gum, and many non-food items (Fig. 3.1). Japanese researchers have identified hydrolases from *Saccharomyces, Alginomonas, Trichoderma* and various bacteria that preferably cleave L-menthyl acetate, chloro acetate, succinate, dodecanoate, and further L-menthyl esters in racemic mixtures. Thus, the hydrolases discriminated equatorial and axial conformation of the hydroxy substituent of the cyclohexanol derivatives. About one third of the yearly consumption of about 4000 metric tons is produced as nature identical L-menthol. Racemic menthol is synthesized by hydrogenation of thymol. (±)-menthol is then separated from the mixture of the four stereoisomers by distillation. Esterification followed by stereoselective hydrolysis yields L-menthol, and the undesired (+)-isomer can be racemized and recycled. This process demonstrates that an enzymatic approach can outdo any other option even on an industrial scale, particularly if chiral aroma chemicals are involved. Future developments could be directed to the stereoselective hydrogenation of readily available L-menthone from peppermint oil.

Kinetic resolution of enantiomer, 2-methylbranched compounds via esterification and acidolysis were achieved with a lipase from *Candida cylindraceae* (Engel, 1992). Analogous reactions with 2- and 3-hydroxy hexanoates yielded lower enantiomeric excess, but indicated combinations of or competition between various enantiodiscriminating pathways rather than the activity of different enzymes. Implications for the analytical work-up and the assessment of the genuine chirality of plant volatiles were outlined. Racemic, optically active alkanolides and alkenolides, such as jasmolactone or tuberolactone, were separated using a commercial horse liver esterase (Givaudan-Roure, 1994). Relactonization of the enantiomer hydroxy acid yielded both optical forms as pure lactones. It was claimed that nature-identical blends gained sensory quality due to the lack of sensory interference from the other enantiomer (cf. Sect. 3.3) Furthermore, the usual interpretation of the legal definition of a 'nature-identical' aroma is challenged by this work (cf. Sects. 1.3, 3.2). Chiral secondary alkanols were prepared via kinetic resolution using esterification with an achiral acid, and alkanolides via intramolecular trans-esterification (Lutz et al., 1992). These attempts partially depended on reversed hydrolysis and shift the scope to enzyme catalyzed reactions in nonaqueous media (Kazandjian et al., 1986).

7.1.3 Reverse Hydrolysis

Substrates and products that are either insoluble or labile in water can now be considered for enzymatic reactions. In nonaqueous media the equilibrium of hydrolytic reactions can be easily changed to condensation conditions. Usually, some water is added to the organic phase to prevent the enzyme protein from desiccation. A catalytically competent ionic state of the enzyme preserves the original conformation, and it was observed that freeze-dried enzymes 'memorize' the pH of the solution from which they were obtained. Water activity and chemical type of the immobilization matrix were demonstrated to be crucial factors for both enzyme activity and stability. Redox reactions may benefit from organic phase biocatalysis, because the usual cofactors are virtually insoluble in apolar solvents, and the effective concentration of cofactor in the aqueous proximity of the enzyme is probably near saturation. This does not, however, relieve us from coupled cofactor regeneration.

Lipase catalyzed reversed hydrolysis or transesterification have numerous food applications, for example the synthesis of modified triacyglycerols, emulsifiers, peptides, and oligosaccharides (Vulfson, 1993). Some guidelines for the synthesis of a monoterpenol butanoate, pentanoate, or similar flavor esters are as follows:

- Depending on substrate size and functionality several lipases can serve as a good catalyst. Lipase of *Candida cylindracea* has gained some popularity. Food grade considerations play a role. A too crude enzyme preparation will contain a lot of hydrated, nonactive protein that acts as an efficient trap for the ester product. Intermittent hydration of the immobilized lipase may improve the long-term stability.
- The alkyl moiety may constitute the reaction medium. If a solvent is to be chosen, the log P value (P is the partition coefficient of the compound in octanol/water) should be >4. Particularly suitable are longer chain alkanes ($>C_6$) and alkanols ($>C_{10}$) and phthalates.
- Water activity must be adjusted to maintain the active conformation of the enzyme according to water saturation of the organic phase. Water produced during the synthetic reaction should be controlled (cold finger, molecular sieve, azeotropic distillation) to avoid a stable emulsion that can make recovery difficult.
- The molar ratio of alkyl/acyl moiety is usually found optimal in the range from 4 to 10, like for chemosynthesis.
- pH optima of synthesis and hydrolysis are usually very similar and in the range from pH 5.0 to 8.0.
- Transesterification temperature can be raised up to 100 °C, because of the reduced thermoinactivation the absence of excess water, particularly of immobilized lipase. Volatile esters may be formed at ambient temperature, but at

reduced conversion rate. Incubation at 37 °C under modest agitation is a good compromise.

Esters of acids from acetic to hexanoic and alcohols from methanol to hexanal, citronellol and geraniol were synthesized by lipases from *Mucor miehei, Aspergillus* sp., *Candida rugosa,* and *Rhizopus arrhizus* (Langrand et al., 1990). Good yields of methyl and ethyl acetate were obtained after the reaction medium *n*-heptane was diluted with diethyl ether to make up 30% (v/v). Ethyl acetate is an important top note in many fruit aromas. Methylbutanoate and methylbutyl esters, key components in apple, banana, and pineapple aroma, are as well amenable by *Candida* lipases. The same catalyst provided the chiron butyl-(3S)-hydroxybutanoate by enantioselective esterification of the racemic acid. (Chattopadhyay and Mamdapur, 1993). With toluene as the reaction medium the enantioselection was excellent for the esterification step and moderate for the kinetic resolution of the ester. Transesterification reactions was made irreversible using vinyl ester substrates (Miyazawa et al., 1992). This was applied to resolve racemic 2-phenoxy propanoic acids. Both the organic reaction medium and the nucleophilic moiety affected enantioselectivity. When chiral alkyl and acyl moieties were esterified, the lipase of *Candida cylindracea* exhibited selectivities toward both of the racemic substrates (Chen et al., 1993). This doubly enantioselective reaction proceeded in *n*-hexane.

7.1.4 Nonaqueous Reaction Media

The optimization of the reaction medium is still a matter of trial and error. A transesterification between glycerol tributanoate and 2-pentanol in iso-octane was catalyzed by a lipase of *Candida rugosa* (Triantafyllou et al., 1993). Increased yield and selectivity was found upon addition of water (0.05% v/v) and *N,N*-dimethylformamide (0.2% v/v). Using the same solvent, the surfactant sodium (bis-2-ethylhexyl)-sulfosuccinate stabilized a microemulsion containing a lipase of *Penicillium simplicissimum* (Stamatis et al., 1993). The enzyme catalyzed the stereospecific esterification of menthol with fatty acids. (+), (-), and *rac*-menthol were studied to establish a fast system for the resolution of racemate according to the approach mentioned above. (1R)- and (1S)-cyclohex-2-enols were prepared by an enantioselective transesterification of a racemic phenyl seleno derivative (Izumi et al., 1993). Before a flavor application of such a process can be realized, the required chemical steps need to be replaced by enzyme technology. The cyclohexenols, however, are useful chiral synthons. A series of (racemic) cyclohexyl esters occur as artificial flavorings on the FEMA list (cf. Sect. 1.3).

Esters of thio and dithio fatty acids were found in hop oil, cheese, other fermented food, and even in meat and fruits. In view of the activation of acyl moieties as coenzyme A thio esters in living cells, it is not surprising to find lipases catalyzing stereoselective hydrolysis and esterification of some of the thio analogues of the above mentioned compounds. Recent experimental work on sulfhydryl reagents has demonstrated, however, that bipolar thio compounds inhibited human pancreatic lipase non-competitively (Gargouri et al., 1992). Modified incubation systems will be required to extend the range of flavor substrates.

7.1.5 Immobilization

Since the beginnings of nonaqueous lipase technology, immobilization was one of the key parameters of success. Starting with adsorption onto glass, other supports, such as porous glass, kieselghur, alumina oxide, and agar beads have been investigated comparatively (Cao et al., 1992). Kieselghur and agar beads gave the highest catalytic activities of all lipases examined. A membrane reactor with chemically immobilized lipase was reported for the esterification of *n*-octanoic acid with racemic 2-butanol (Ceynowa and Sionkowska, 1993). The activity of the membrane bound lipase was about twice as high as that of the native enzyme, and an ee of <85% of the produced ester mixture was found. Very volatile flavor esters were synthesized in a packed-bed gas phase reactor (Parvaresh et al., 1992). A crude porcine pancreatic lipase was sandwiched between two layers of dehydrated glass wool. Substrate was added by passing a nitrogen stream as a carrier through two flasks containing an alcohol and an ester, and then through the thermostatted column reactor. The transesterified product was solvent trapped. Water activity of the enzyme was essential. Improved yields can be expected from an optimized process regarding the elevated temperature (60 °C) and diffusion rates in the gas/solid-reactor.

7.2 Carbohydrases

Like other hydroxy metabolites of plants, volatile compounds can be acetalized to better water soluble transport and storage conjugates. Years after the first positive identification of geranyl-β-D-glucoside in rose flowers (Francis and Allcock, 1969), bound forms of volatile flavors have started to attract interest as nonvolatile flavor precursors.

7.2.1 Glycosidic Aroma Precursors

Two recent conferences have devoted individual sessions to glycosides and related polyol compounds (Teranishi et al., 1992; Schreier and Winterhalter, 1993). Work carried out in the 1990s has been summarized in Table 7.2. Usually reported as aglycons were medium chain alkanols and alkenols (including (Z3)-hexenol), monoterpenols, phenolics, and phenyl substituted alcohols. Products of carotenoid degradation were frequently isolated from grape and wine. Bound ionones and ionols were identified, among others, in extracts of nectarine, raspberry, and starfruit, while bound 2,5-dimethyl-4-hydroxy-2*H*-furan-3-one occurred in pineapple

Table 7.2. Glycosidically bound volatiles in plants

Source	Reference
Apricot	Salles et al., 1991
Grape skin and mesocap	Park et al., 1991
Grape juices	Francis et a., 1992
Grape and wine	Sefton et al., 1992
Grape cvs.	Razungles et al., 1993
Lulo *(Solanum vestissimum)*	Wintoch et al., 1993
Mango	Adedeji et al., 1992
Mango	Koulibaly et al., 1992
Nectarine fruit and juice	Takeoka et al., 1992
Passionfruit, apricot, mango	Challier et al., 1990
Pineapple	Wu et al., 1991
Raspberry	Pabst et al., 1991
Sloe *(Prunus spinosa)*	Humpf and Schreier, 1992
Starfruit *(Averrhoa carambola)*	Herderich et al., 1992
Strawberry	Wintoch et al., 1991
Tea (Oolong cv. Shuixian)	Guo et al., 1993
Tomato	Marlatt et al., 1992
Vanilla bean	Ranadive, 1992
Wine (cv. Riesling)	Marinos et al., 1992
Wine and others	Williams et al., 1993
Eriobotrya japonica	De Tommasi et al., 1992
Hovenia dulsis leaf	Yoshikawa et al., 1993
Jasminum, Gardenia flower	Watanabe et al., 1993
Nicotiana sp. flower	Loughrin et al., 1992
Osmanthus asiaticus bark	Sugiyama et al., 1993
Ribes uva crispa, Sorbus aria leaves	Humpf et al., 1992
Scutellaria orientalis leaf	Calis et al., 1993
Vitis vinifera leaf	Roscher and Winterhalter,1993

and strawberry (Fig. 7.1). Other volatiles that are clearly unable to form glycosidic bonds with a carbohydrate moiety have regularly been reported.

Less attention has been paid to the glycon part of the molecules. Crude pectinase preparations with known β-glucosidase side activity or purified β-glucosidases were incubated with the nonvolatile flavor fractions to yield volatile products upon hydrolysis. In fact, β-D-glucosides appear to constitute a major part of the glycosidic pool of many plants, but other glycons, such as pentoses, methylpentoses,

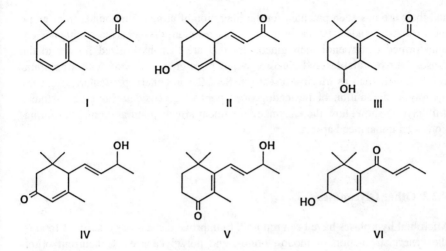

Fig. 7.1. C_{13}-norisoprenoids of the glycosidic fraction of raspberry. I = 3,4-Didehydro-β-ionone, II = 3-Hydroxy-α-ionone, III = 4-Hydroxy-β-ionone, IV = 3-Oxo-α-ionol, V = 4-Oxo-β-ionol, VI = 3-Hydroxy-β-damascenone (Pabst et al., 1991)

other hexoses, and di- and trisaccharides have also been reported (Fig. 7.2). Precursors of *Gardenia jasminoides* volatiles were suggested as being not mainly β-glucosides, because a naringinase treatment was much more efficient than almond β-glucosidase (Watanabe et al., 1993). Further genuine moieties, such as phosphate or malate, usually remain undetected due to hydrolytic side activities of the crude enzyme preparations used.

With this biochemical background considerations were focussed on the application of glycanases to intensify the aroma of food products. In wine, the grape β-glucosidase is rapidly inactivated during centrifugation/pasteurization of the must or during fermentation. Glucosidases of *S. cerevisiae* and of *Candida molischiana* were suggested as substitutes (Vasserot et al., 1993). However, many fungal glycosidases are inhibited by glucose, fructose, and ethanol, and display low activity at the pH of wine. Some β-apiosidases, α-arabinosidases, and α-rhamnosidases of strains of *Aspergillus niger* do not suffer from these drawbacks (Dupin et al., 1992). The formation of these enzymes was induced by the respective glycosides,

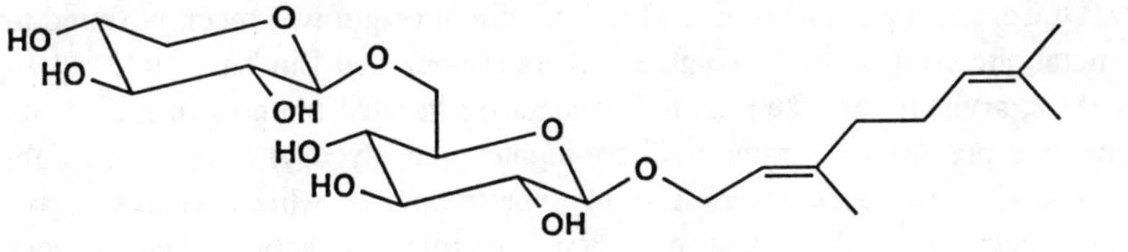

Fig. 7.2. Geranyl 6-*O*-β-D-xylopyranosyl-β-D-glucopyranoside from *Oolong* tea leaves (Gao et al., 1992)

and their use has been patented for the liberation of aroma compounds from grape must (Gist-Brocades, 1991). In a sequential reaction, disaccharide conjugates are transformed to terpenol monoglucosides that are then hydrolized further to the sensory active volatiles and glucose. Wines from grape cultivars with a pronounced bouquet, such as Morio-Muscat (cf. Sect. 2.3.1) benefit particularly from the hydrolytic diminution of the endogenous pool of glycosides. Long-term studies will have to show, how the enzymatic treatment affects maturation and the formation of an imbalanced aroma.

7.2.2 Other Glycanases

Microbial hydrolases have been reported to improve the sensory quality of food by the synergistic action of mono-, oligo-, and polyglycanases. Concurrent observations on increased levels of free vanillin in fermented foods supplemented with vanilla powder or vanilla extracts indicated that certain food species, such as *Lactobacillus bulgaricus* can hydrolyze glycosides not normally occurring in their innate environments (cf. Sect. 5.4.1). A recent patent claimed a process for the production of vanilla extracts involving treating crushed green vanilla beans with enzymes capable of degrading both the plant cell walls and the glucosidic precursor (Pernod-Ricard, 1993). The process reduced the time for producing vanillin and increased the yields. Similarly, the processing of matsutake mushroom was improved by carbohydrases of *Trichoderma viride* (Ebara Sangyo Jugen, 1993). A thermal inactivation step prevented an undesirable transfer into the mushroom filtrate. Cell wall degrading enzymes were combined with anaerobic lactic bacteria to increase fiber digestibility and recovery of cellular metabolites (Tengerdy et al., 1992). The process increased the yields of carotene, chlorophylls and xanthophylls from various materials, and also the recovery of sclareol from clary sage (*Salvia sclarea*). The latter is of interest for the industrial production of ambra fragrances (Fig. 7.3). A glycosidic form of sclareol cannot be excluded from the tertiary structure of its hydroxy functions, as the examples linalool and hotrienol clearly demonstrate (cf. Salles et al., 1991; Park et al., 1991; Wintoch et all., 1993). Pectinases, cellulases, and hemicellulases are now common liquefiers in fruit (juice) processing. Due to their specificities for polymer substrates, they usually leave low molecular weight glycosides untouched. A cellulase contaminated with a glucosidase side activity, however, liberated benzaldehyde during the processing of peach (Di Cesare et al., 1993). Even food of animal origin was recently found to contain metabolic conjugates of volatile flavors (Lopez and Lindsay, 1993). β-D-glucuronidase, arylsulfatase, and acid phosphatase liberated phenols in milk from their respective precursors. Amino acid conjugates and glycosyl esters of straight chain fatty acids have been inferred from experiments in which the respective acids were released by N-acylase and β-D-glucuronidase action. The sensory properties of the volatiles liberated were responsible for the typical differences of the odor of cow and sheep milk.

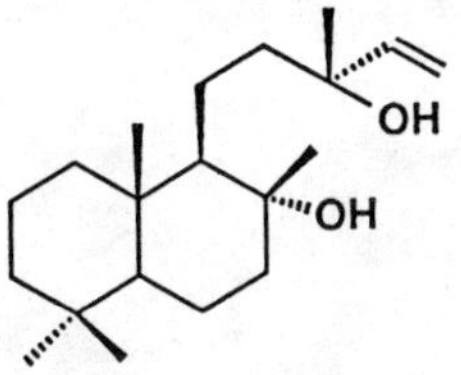

Fig. 7.3. Sclareol, a labdanoid diterpendiol, is probably glycosidically bound in *Salvia sclarea* (Tengerdy et al., 1992)

Various carbohydrases have recently been purified and characterized, for example a β-glucosidase from *A. niger* (Unno et al., 1993), and an acid α-glucosidase with maltase properties from banana pulp (Konishi et al., 1992). Carbohydrases have contributed to the assessment of identity and origin of plant products, to our understanding of the development during processing and maturation, and to the selection of aroma-rich cultivars. Stability and selectivity data will be decisive for the sensory changes of a food and, thus, for the future application of the new enzymes in food processing. Moreover, glycosidic forms of volatiles are becoming more and more recognized as chemically and physically stable, water-soluble flavor depots: For example, phenyl-β-glucoside as a model compound was added to cornmeal prior to extrusion; phenol was produced from its nonvolatile precursor during thermal processing (Tanaka et al., 1994). As the specific chemo-synthesis of glycosides inevitably requires multiple protection/deprotection steps, a biotechnological transglycosilation deserves attention. Microbial glycosyltransferase have been characterized (Park et al., 1992). Selective transglycosylations can be achieved using glycosidases. Lactase and amylase catalyzed transglucosylation reactions in an aqueous phase (Matsui et al., 1994). The demand for stable precursors is illustrated by synthetic acetal derivatives for aroma aldehydes, such as benzaldehyde or cinnamaldehyde (Anderson, 1993). The aldehydes or their acetals were reacted with either methyl-β-D-glucopyranoside to 4,6-glucoside acetals, or with dialkyl tartrates or 2-hydroxyacids to the respective cyclic acetals (Fig. 7.4). These odorless derivatives, though artificial, provide an interesting alternative to encapsulation and are easy to transport and formulate. To avoid toxicological problems, tartrate, lactate, and other food constitutents were selected as hydroxy agents.

Another carbohydrase, naringinase, is involved in a biotechnological process for the production of 2,5-dimethyl-4-hydroxy-3(2*H*)-furan-3-one (cf. Fig. 1.1). Citrus wastes contain the flavanoid glycosides naringin and rutin. Hydrolytic cleavage yields a rhamnose/glucose/naringenin mixture, from which the aglycon is easily removed. A complete and efficient separation of glucose from the remaining rhamnose can be achieved by either selective fermentation, or by glucose oxidase catalyzed oxidation, or by fermentation to 5-oxogluconic acid using *Gluconobacter*. Rhamnose is finally purified by crystallization and reacted to the target compound in the presence of an amino acid (Cheetham, 1992).

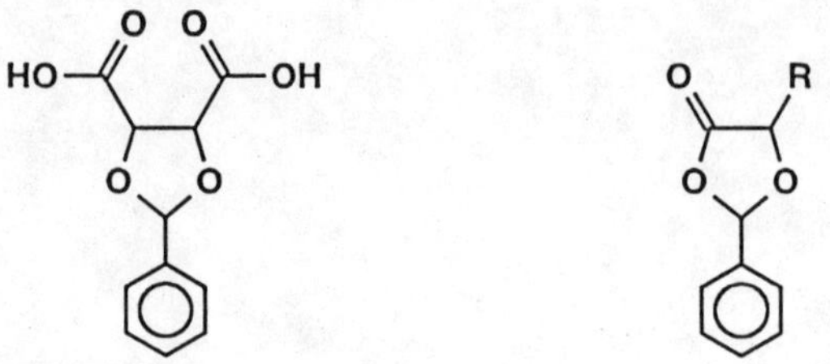

Fig. 7.4. Tartrate and 2-hydroxyacid derived acetals imitating natural conjugates of aromatic flavour aldehydes (Anderson, 1993)

7.3 Proteases

Smaller peptides and free amino acids, the end products of the activity of the various proteases, contribute to the nonvolatile flavor fraction and act as precursors for volatiles. A salt-resistant glutaminase was patented to enrich pickled cabbage in 'taste and flavor' (Daiwa Kasei, 1991). A combination of a heat-stable protease from a *Bacillus* sp. and ultrafiltration membranes was described to produce a natural seasoning from a fish cooking waste stream (Ishikawa et al., 1993). By analogy to lipases and carbohydrases, the hydrolysis could be partially reversed for certain enzymes under appropriate conditions. A *Pseudomonas aeruginosa* elastase catalyzed amide formation of *N*-blocked-L-aspartic acid and *C*-blocked L-phenyl alanine to yield an aspartame precursor (Lee et al., 1992). Depending on reaction temperature various concentrations of methanol as a cosolvent accelerated the aqueous reaction; hence, measures to shift the reaction equilibria appear promising.

The structures of α-chymotrypsin and horse liver alcohol dehydrogenase were compared in isooctane reversed micelle solutions (Creagh et al., 1993). Large structural perturbations accompanied the solubilization of the dehydrogenase, while the structure and activity of the protease was not adversely affected. The authors compared the structural integrity of both enzymes with the catalytic activity in reverse micelles, and concluded a direct correlation that was not dependent on water activity of the system. The conformation of redox enzymes can be likewise stabilized by immobilization or by chemical modification to open applications in the field of aroma compounds.

7.4 Oxidoreductases

For those who enjoy classification of terms, three major categories of enantiodifferentiation by enzymes may be distinguished:

- Enantiomer differentiation, for example the lipase catalyzed kinetic resolution of racemates as discussed above for D,L-menthyl esters (cf. Sect. 7.1.2).
- Enantiotope differentiation, the transformation of prochiral functions (or *meso-*

forms) into chiral centers. The substrate bears two chemically identical, but ste-reochemically different enantiotops (cf. Sect. 6.1).
- Enantioface differentiation, the transformation of a planar, achiral center into a chiral center (cf. Sect. 6.3, Fig. 6.11).

Numerous examples show that hydrolases differentiate substrates according to all of the three categories (Faber, 1992). The situation is more complicated with redox enzymes: Dehydrogenases were used to separate racemates of secondary alcohols, but this enantiomer differentiation ultimately based on a two-step enantioface differentiation. Specific enantioface reductions of carbon-oxygen and carbon-carbon (enoate reductase) bonds have obvious flavor applications and have fre-quently been approached using whole cells. The enantiotope asymmetrization of *meso*-diols has been used to generate 4-and 5-lactones. Monooxygenases recogni-ze enantiotops, for example *exo/endo* hydrogens of cyclic monoterpenes, such as 1,4-cineol. Another flavor application, the oxidation of ketones to esters or lacto-nes in a Baeyer-Villiger type enzymatic reaction proceeds under chiral recognition according to category one. Dioxygenases, such as the above mentioned lipoxyge-nases (cf. Sects. 1.5, 5.3.2), differentiate enantio faces of aliphatic and aromatic substrates. In spite of the commercial success of glucose oxidase, peroxidases or other oxidases have found very limited applications in aroma biotechnology.

Horse liver alcohol dehydrogenase with its broad substrate specificity, narrow stereospecificity, and bidirectional functionality has been most widely used in the transformation of aroma compounds, but there is continuing research for novel reductases. A recent study screened 65 microorganisms to find active keto ester reductases (Peters et al., 1992). The specific activity of reductases of *Candida parapsilosis* and *Rhodococcus erythropolis* were enhanced by growth on *n*-alkanes. An intermediary keto ester was postulated as an inducer. Enantiomer 3-hydroxy acid and 3-acetoxy acid esters occur in the aroma of many fruits. Cyto-chrome type monooxygenases generate the activated oxygen transferring species by an iron containing protoheme cofactor, while the flavin dependent enzymes form a heteroaromatic hydroperoxide species. Much of our knowledge on cyto-chrome P-450 enzymes was derived from the D-(+)-camphor inducable system of *Pseudomonas putida* (Horowitz and Vilker, 1993).

Fed-batch and continuous cultures were optimized for cytochrome production. Key parameters were elimination of iron deficiency, oxygen enrichment, and si-multaneous decrease of total gas throughput. The camphor solvent, dimethylfor-mamide, was assessed for its inhibitory effect. Numerous insertions of oxygen into precursor substrates can yield odorous alcohols, epoxide derived compounds, or esters. Hydroxylations of monoterpenes were comprehensively reviewed (Karp and Croteau, 1988).

The intriguing oxidation of (bio)ethanol to the flavor impact acetaldehyde has been adressed above (cf. Sect. 5.3.3). Instead of using whole cells, alcohol oxida-se, an unidirectional flavine enzyme, was immobilized on Amberlite IRA-400, packed into a column reactor, and the production of acetaldehyde performed in the gas phase (Hwang et al., 1993). Reactor temperature and input flow rates affected the bioconversion.

7.5 Miscellaneous Enzymes

Other activities encountered in whole cell transformation reactions have been attempted to isolate and utilize in purified form. Considering the formation of flavor esters by sake yeast (cf. Sect. 2.9.2) the alcohol acetyltransferase was characterized (Minetoki, 1992). Cinnamoyl esterase and cinnamate decarboxylase may be prepared from commercial crude enzymes to produce 4-vinylphenols with impact attributes (Dugelay et al., 1993). Lyase activity of a *Pseudomonas putida* methionase was described to hydrogen sulfide and methanethiol, key odorants in cheese and other food (Lindsay and Rippe, 1986). Reactions carried out under anaerobic conditions yielded mainly methane thiol, while aerobic conditions favored further oxidation (Fig. 7.5). Incorporation of fat encapsulated methionine/methionase into cheese accelerated flavor development. Enzymatic formation of new carbon-carbon bonds has been rarely used in aroma biotechnology. Michael type additions occurred during the bioconversion of myrcene by basidiomycetes (cf. Sect. 6.1). An aldolase catalyzed reaction was involved in an often cited sequence leading to 2,5-dimethyl-4-hydroxy-3(2H)-furan-3-one (Wong et al., 1983). A TPP-linked benzoylformate decarboxylase of *Ps. putida* produced 2-hydroxypropiophenone from benzoylformate and acetaldehyde (Wilcocks et al., 1992). Oxidation of this acyloin would yield a 2,3-diketone that was reported as a constituent of coffee aroma. Market demands and chemosynthetic access will determine the viability of an approach by enzyme technology.

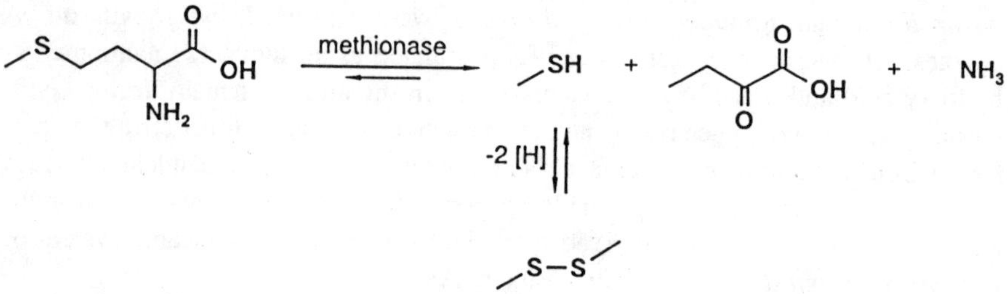

Fig. 7.5. Methionase catalyzed formation of cheese aroma impacts (Lindsay and Rippe, 1986)

7.6 Cofactor Recycling

Cofactors providing chemical energy or transferable chemical agents are required in all redox reactions and often in reactions catalyzed by other classes of enzymes. As a supplementation with stoichiometric amounts of cofactor is usually impossible, the saturation level has to be maintained by a continuous regeneration during the bioprocess. The regeneration of ATP and NAD is economically feasible and is now operated on a scale of metric tons. The coupled-substrate method and the coupled-enzyme method rely on complementary substrates (Faber, 1992) and one or possibly two enzymes sharing the same cofactor. Synthetic reaction chains can be envisaged by assembling appropriate partners (Fig. 7.6). Cross reactivities of the enzymes as well as substrate and product inhibition must be excluded by sophisticated designs. Coimmobilization of the enzymes on a common support will facilitate the shuttle of substrates and cofactors. The terminal donor substrate (A-X) can be continuously fed from an abundant source or, for example, regenerated electrochemically to support a redox chain (Kataoka et al., 1992; Itozawa and Kise, 1993). Several flavor applications have been presented (Berger, 1991). The bienzymatic method was used in coencapsulating enzymes of *Streptococcus lactis* and *Gluconobacter oxidans* with L-leucine, ethanol, and NAD to produce acetic acid and the target compounds 3-methylbutanal and 3-methylbutanol in a cheese related matrix (Braun and Olson, 1986). Such coimmobilisates of catalysts of different origin, or of biocatalyst and its substrate belong to a second generation of immobilized catalysts. They seem promising in the area of aroma production, because poorly water-soluble and volatile substrates must be handled.

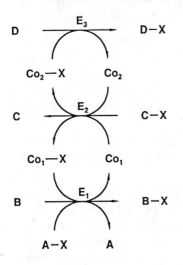

Fig. 7.6. Synthetic electron chain in cofactor dependent enzymatic reactions

8 Genetically Altered Catalysts

Preceding chapters have dealt with de novo syntheses, conversions and transformation reactions of substantial potential for the biotechnology of aromas. Satisfactory yields and the desired target compound(s) have, however, often not been achieved instantly. This has stimulated interest in classically mutagenized and genetically engineered strains with improved performance. The recombinant DNA technique has dramatically extended possibilities by providing a rapid means of introducing predetermined genetic alterations with high accuracy. A more rapid progress in the field is still impeded by safety considerations, public concern, and legal and administrative (partially self-) restrictions. Instead of completely covering all aspects of the emotional discussion, some facts may be remembered:

- Gene transfer is an everyday event in nature. It was even hypothesized that bacterial transformation evolved as a nutrient uptake mechanism, because unrelated DNA is abundant in some natural environments (Redfield, 1993).
- Plasmid harboring cells are ecologically handicapped as compared to the wild strain. Difference in growth rates has been suggested as a diagnostic character (Park and Ryu, 1992).
- Safety testing of food generated with the aid of genetically manipulated strains is emphasized, particularly if the microorganisms are to remain in the consumed product.

A recent review stated that no example would yet exist for 'pathway manipulation of a fragrance or flavor chemical'. Though not all of the following notes will directly refer to a concerted bioprocess for an aroma compound, it will become clear that active pathway manipulation has been demonstrated to overcome some of the present problems. Passively 'pathway manipulated' or actively mutagenized food strains are being industrially used since decades anyway.

8.1 Mutants and Fusants

The malolactic transformation is a desirable trait of *Leuconostoc* or *Lactobacillus* during white wine fermentation (cf. Sect. 2.3.4). A lack of this ability is desirable during the fermentation of brined vegetables, because the developing carbon dioxides contributes to the formation of bloaters. Mutants unable to perform the decarboxylation reaction were selected (mdc⁻ phenotype), and a low reversion frequency was reported (Breidt and Fleming, 1992). The acceleration of cheese ripening by increasing the amount of starter culture is limited by the rate of acid production. Procedures for the generation and selection of mutants of lactic acid bacteria are being widely employed for the production of improved starter cultures. A novel approach was the use of lac⁻ mutants of *Lactococcus lactis* in addition to the regular starter (Birkeland et al., 1992). A more rapid development of soluble nitrogen compounds and higher flavor scores of experimental cheeses were reported. High total starter populations of *Streptococcus lactis* lac⁻ prt⁻ double mutants had similar effects. Product inhibition, a key problem of biotechnology, was overcome for lactic acid by using a mutant of *Lactobacillus delbrueckii* that tolerated increased acid levels (Demirci and Pometto, 1992). A significant increase in productivity was the result.

Classical hybridization or mutagenesis and selection of a desired phenotype of industrial yeasts is hampered by the polyploid or aneuploid nature of these genetically highly stable strains. Some success has been reported for the generation of primary alcohols, particularly 2-methylpropanol, the isomer methylbutanols, and 2-phenyl ethanol that contribute significantly to the aroma of yeast fermented foods (cf. Sect. 2.3). As their biogenesis is known, attempts were made to select clonal variants or conjugational mutants, for example with an altered amino acid metabolism (Akita, 1992). It has been argued whether novel 'banana' or 'Bartlett pear' odor attributes of Beaujolais wines were derived from 'aroma designed' yeast mutants. A diploid yeast mutant was ergosterol auxotroph and released monoterpenols into the medium (Javelot et al., 1991). Fermented grape musts, as a result, received muscatel odor attributes. A simple selection protocol was described for sake brewing (Watanabe et al., 1993): Yeasts were cultured in the presence of 3-methylbutyl monofluoroacetate, and inhibition by the liberated free acid was observed. Mutants with low esterase activity survived. These strains did not attack their own ester products to the same extent as the standard yeast. Increased levels of isoalkyl acetates were accumulated and imparted a fruity aroma in sake. The inverse approach was used selecting an ester overproducing mutant of *Saccharomyces cerevisiae* from 925 strains (Saito et al., 1992). The brewed sake contained a large amount of ethyl hexanoate, a strong fruity odorant. An *N*-methyl-*N'*-nitro-*N*-nitrosoguanidine treated *Saccharomyces* produced an intense flavor caused by markedly increased formation of fusel alcohol (Pan and Kuo, 1993). Some overproducers were metabolically characterized in more detail. Haploid, low acid and high ester producing yeast strains were mated for sake brewing (Yoshida et al., 1993). The resulting strain combined the positive sensory attributes. A wine yeast was crossed with an ergosterol auxotroph of *Saccharomyces cerevisiae* that accumulated monoterpenols (Pernod-Ricard, 1992). Meiotic descendants of this cross produced the same terpenols, thereby imparting a muscat note to fermented beverages. Hence, the above mentioned lack of volatile terpene synthesis of yeast was overcome (cf. Sect. 2.3.1). The enhanced production of higher alcohols of a haploid *Zygosaccharomyces rouxii* was related to a defective uptake system for L-leucine and L-phenyl alanine (Aoki and Uchida, 1991). A *Saccharomyces cerevisiae* mutant had a threonine deaminase (EC 4.2.1.16) with a decreased feedback sensitivity to L-isoleucine and accumulated this amino acid and the respective intermediary alcohols (Fukuda et al., 1993). Sake and bread (Watanabe et al., 1990) made with the mutant yeast contained more volatile components and had an enhanced flavor. Bread containing more methylpropanol received higher flavor scores, while methylbutanol, by contrast to sake, caused unpleasant attributes. 2-phenyl ethanol and some of its derivatives possess rose and honeylike odors that round off the flavor of alcoholic beverages, other foods, and cosmetics. These odorants have been accumulated in yeast mutants (Fukuda, 1992). *Zygosaccharomyces rouxii* mutants were selected according to their resistance toward *p*-fluorophenyl alanine (Yoshikawa et al., 1992). A mutant, deficient in prephenate dehydrogenase, produced high levels of 2-phenyl ethanol and was suggested for use in traditional fermentations. Earlier work of Fukuda's group (1990) has prepared the ground for this type of study.

If desired metabolic traits are distributed over different strains (cf. Sects. 2.10, 5.5), intra- or intergenetic fusion of protoplasts offer a solution to overcome the poor sporulation and mating abilities of industrial yeast strains. An improvement can be obtained either technically, for example thermotolerance and flocculation (Kida et al., 1992), or metabolically. Protoplast fusion of wine (*S. cerevisiae*) and soysauce yeast (*Z. rouxii*) has been successfully attempted (Kusumegi et al., 1992). The fusant grew well and continued to produce the typical aroma components. Haploid strains of *Aspergillus niger* were fused to increase the production of a β-glucosidase (Hoh et al., 1992). Implications of glycosidic enzymes in flavor generation were discussed in Sects. 7.2.1. and 7.2.2. Methods for effectively fusing cells of higher fungi, including basidiomycetes, were also developed (Kawasumi, 1989). A mutagenetic treatment of the protoplast prior to fusion opens the way to strains with improved aroma formation capabilities (cf. Sect. 5.4). Problems encountered are the choice of selection markers, the loss of specific parental qualities, and the often insufficient genome stability of the fusants.

8.2 In Vitro rDNA Technique

The development of methodologies to isolate and purify microbial DNA, coupled to the discovery of various vectors that are capable of transferring and replicating foreign DNA in a host organism, has allowed many genes to be characterized on a molecular level. Inverting the direction of transfer, these genes were also returned to their donors to determine the effects of modifying their structure or their expression. Food technologists have designed applications for both the transformed microorganism and the primary products of gene expression, the enzyme proteins. The advantages of a new enzyme may be seen in an extended range of usable substrates or in an improved substrate transport, and the enzyme as such can be the bioprocess target. Secondary products of gene activity usually result from enzymatic chains and are much more difficult to obtain from a recombinant cell. This likewise refers to trophophase and idiophase metabolites (cf. Sect. 5.6), and includes, thus, also aroma compounds.

8.2.1 Enzymes

Chymosin, the traditional milk clotting agent, is an aspartyl protease (EC 3.4.23.4) that is extracted from the fourth stomach of suckling calves. With increased production of cheese, sufficient supply from the animal source was getting more and more difficult. As a result, the calf gene was incorporated into *Kluyveromyces lactis*, and the secreted protein became the first enzyme product from an industrial scale cultivation of a recombinant strain. Chymosin makes a significant contribution to the initial stages of (αs1-) casein degradation, thereby delivering key precursors of cheese flavor.

Evidence that tailor-made enzymes will be needed in the future comes from the numerous reports on modifying soluble forms by immobilization or chemical derivatization. Cloning and expressing foreign enzymes in (preferably food grade) recombinant cells would be a primrose path. Current genetic approaches revolve mainly around lipases. DNA fragments encoding for lipases of *Rhizopus niveus* (Kugimiya eta al., 1992) and of *Penicillium camembertii* (Yamaguchi et al., 1991) have been cloned and sequenced. The tag-lipases of three different fungal species showed remarkable homologies of the primary structure including conservation of the catalytic triad (cf. Sect. 7.1). The sequence of a gene encoding for a thermostable lipase of *Pseudomonas fluorescens* was determined, and an 1347 bp open reading frame was characterized (Chung et al., 1991). Enabling an elevated reaction temperature, such an enzyme could be used to transesterify higher boiling moieties of aroma esters. On the other hand, sensitive or highly volatile moieties should be reacted at a lower temperature. Lipase encoding genes of an antarctic species were introduced into *Escherichia coli* to achieve a significant drop in the optimal reaction temperature (Feller et al., 1991). *Staphylococcus carnosus*, used as a starter in salami manufacture, was successfully transformed to secrete a heterologous lipase or other proteins (Götz, 1990). Strong promotors can be obtained and cloned from genes of strains that are known to secrete large amounts of hydrolases. Synthesis of aroma esters is not limited to lipases, as was demonstrated using a recombinant cutinase of *Fusarium solani* (Sebastiano et al., 1993). Maximum specific activity was observed with *n*-hexanol. This substrate also proved to be a strong stabilizer of the enzyme.

Medium chain aldehydes and carboxylic acids constitute large groups of aroma chemicals and flavor building blocks. As the corresponding alcohols are often available, the dehydrogenase enzymes are of industrial interest. An alcohol dehydrogenase subunit encoding gene from *Acetobacter aceti* (Inoue et al., 1989) and an aldehyde dehydrogenase encoding gene from *Aspergillus niger* (O'Connell and Kelly, 1989) were sequenced. Homologies existed with corresponding and related genes of different origins. Using these enzymes in continuously operated membrane reactors offered a convenient route to many fruity or 'green' key odorants; however, cofactor demand must be met (cf. Sect. 7.6). Similar considerations apply to the genes of carotenogenesis identified in *Erwinia herbicola* (Schnurr et al., 1991). Six genes, among them a gene encoding a β-carotene hydroxylase, were identified on a chromosomal fragment. The incorporation into a food strain could permit access to many valuable norisoprenoid volatiles, such as the ionones or methylheptenones (cf. Fig. 7.1).

Saccharomyces cerevisiae is a particularly attractive host for the production of heterologous proteins. Unlike prokaryotic food species, its eukaryotic subcellular organization enables carrying out posttranslational processing steps required to generate active eukaryotic enzymes. The GRAS-status (cf. Sect. 1.3) may facilitate approval of genetically engineered strains by the regulatory agencies. Thus, an improved knowledge of the genes of fatty acid synthesis, for example, may have future impact in aroma biotechnology, because so many volatiles are derived from fatty acid precursors (Köttig et al., 1991; cf. Sects. 5.3, 6.3). This applies to the aroma of fermented food and others. Recent reviews have discussed the importance of lipoxygenase started fatty acid catabolism in the development of fruit, vegetable, and mushroom flavor, and have suggested supplementing appropriate raw materials with the respective gene. When a pea seed lipoxygenase cDNA was implemented in a yeast, an effect of the hydroperoxy products on the rheology of a fermented wheat flour dough was expected (Casey et al., 1989). However, further enzymatic activities are required to complete the pathway from hydroperoxides to volatile degradation products (cf. Sect. 5.3.2). More promising for an immediate application is a gene, isolated from a food strain and encoding for a sulfhydryl oxidase activity that could be incorporated into host organisms for the treatment of doughs and the removal of thiols from beer (Quest, 1993). The enzyme was isolated from a commercial preparation of *Aspergillus niger* glucose oxidase, and a recombinant host overproducing the enzyme was constructed.

8.2.2 Lactococci

Lactobacilli, as used in milk processing, are expected to proliferate rapidly and to acidify the growth medium reproducibly. These abilities imply active lactose transport and hydrolysis genes, and a variety of genes encoding for proteases. Surprisingly, the genes encoding for these and other enzymes of primary technological importance occur on independently replicating, labile extrachromosomal elements, the plasmids (Table 8.1). Number (from one to more than 10) and size (from 1 MDa to >100 MDa) of plasmids in a lactococcal strain may differ. Plasmids may be classified as self-transmissible by natural conjugation, nontransmissible but mobilizable by conjugative plasmids, or insertable into the chromosome (Elalami et al., 1992a). In mixed batch cultures of *Lactococcus lactis*, only the self-transmissible plasmid was transferred. Transmission frequencies were about two orders of magnitude lower in an unstirred broth, and five orders of magnitude lower than on agar medium. The arbitrary character of a protoplast fusion or a conjugational gene transfer is overcome by electroporation (cf. chapter 12). High-intensity (6000 Vx cm^{-1}), but short electrical pulses concentrate the like charges on opposite sides of the cell. Transient formation of pores then permeabilizes the cell envelope and facilitates the uptake of exogenous matter. Transferred genetic elements, however, may be lost again during subcul-

Table 8.1. Plasmid located, technologically essential traits of lactococci

---

- Lactose transport and hydrolysis
- β-casein protease, maturation proteins
- Citrate permease and diacetyl formation
- Bacteriocin production
- Bacteriophage resistance

---

tivation, if the external selection pressure is removed, or due to incompatibility (Elalami et al., 1992b). Plasmids carrying insertion elements flanking the replication region can be stabilized by chromosomal integration due to homologous insertion sequences (Jahns et al., 1991). Thus, the development of functional integration vectors could help to stably express important, previously plasmid-located metabolic traits. Transposons as insertion mutagens also provided a useful tool for the localization of gene loci (Israelsen and Hansen, 1993; Romero and Klaenhammer, 1993).

Numerous genes from lactococci, for example, *lacG, prtP, prtM, nisA,* and phage resistance genes, were isolated and cloned. Among them, only the 2-acetolactate decarboxylase gene is directly related to the sensory quality of fermented foods. The enzyme, α-acetolactate decarboxylase, catalyzes the key reaction to acetoin (cf. Fig. 2.2). Its enhancement (alcoholic fermentations) or suppression may be desirable in the respective chemical environments (cf. Sects. 2.2, 2.5). Diacetyl levels in dairy products could be raised by blocking the decarboxylase gene. Alternatively, metabolic engineering could aim at stimulating further steps to α-acetolactate or at improving the active permeation of citrate, the diacetyl precursor (Sesma et al., 1990).

For several years, the heterologous expression of the thaumatin encoding genes of *Thaumatococcus daniellii* in microorganisms as a first example of a plant gene has been tried. The fruits of the plant contain extremly potent proteinaceous sweeteners that were suggested for chewing gum and dairy products. Work to express the gene in *Lactococcus lactis* is continuously pursued (van de Guchte et al., 1992).

The proteinase system of lactococci is indirectly associated with the formation of volatile flavors. The multienzyme complex consists of various proteinases, peptidases, and dipeptide and oligopeptide transport systems. Because of their vital importance for cell growth, they have been genetically studied in some detail (Kok, 1993). Cheeses made with prt⁻ starters lacked flavor (Law et al., 1993). Proteinases encoded by plasmids of *Lactococcus lactis* ssp. were required for the accumulation of small peptides and free amino acids, the cell wall bound proteinases being more active than those secreted into the medium. Mapping a plasmid of a *Lactococcus lactis* ssp., Requena and McKay (1993) found both proteinase activity and lactose utilization located on the same 40 MDa plasmid, while in another ssp, lactose, citrate, and proteinase metabolisms were encoded by three plasmids of different size (Nakamura et al., 1992).

Pyruvate decarboxylase and aldehyde dehydrogenase genes of *Zymomonas mobilis* were transferred to determine the suitability of lactobacilli as a potential host for ethanol producing genes (Gold et al., 1992). *Lactobacillus casei* was harboring a cholesterol oxidase gene of *Streptomyces* and expressed the foreign gene with no indication of deletion mutational events (Somkuti et al., 1992). Sequencing and detection of structural genes of lactococci has taken advantage from the quick amplification enabled by the polymerase chain reaction (Devos et al., 1993). High copy number plasmids were constructed to shuttle genes between lactococci and *E.coli* (Wells et al., 1993), including the malolactic enzyme gene (Chagnaud et al., 1992). Another study bearing potential aroma applications transferred genes of bacterial synthesis of L-phenyl alanine into a *Corynebacterium glutamicum* (Ikeda et al., 1993). A mutated plasmid derivative mediated resistance to end product inhibition and resulted in yields of 23 g×L L-phenyl alanine, a precursor of volatile flavors, such as phenylethyl esters and phenyl acetaldehyde.

The development of integrating vectors in often required for stable expression and for the study of regulation phenomena (Mollet et al., 1993). This approach has already been patented in combination with a deletion of the replication site of the plasmid to block autonomous replication (Unilever, 1992). This type of research is directed toward the construction of 'food grade vectors' without resort to foreign genetic material, with controllable amplification, and with lactococcal metabolic selection markers. Recombinant lactococci bearing enhanced phage resistance, the technologically most important trait, are already used in commercial cheese making (Klaenhammer, 1991). This success will stimulate further molecular genetic studies to construct new starter strains with pronounced or even new flavor formation properties.

8.2.3 Yeasts

Saccharomyces cerevisiae, the best characterized eukaryotic organism, 'acts as a model for eukaryotes in general' (Hadfield et al., 1993), and, meanwhile, rivals the established methodology around *E. coli* with respect to:

– cell permeabilization techniques (virtually fusion techniques),
– appropiate vector constructs,
– selectable (metabolic) markers, and
– auxotrophic (Leu, His, Trp) recipient strains.

Integrating vectors, constructed of bacterial sequences for replication and genetic selection, and a yeast selectable marker are used for obtaining stably altered phenotypes. Episomal vectors carry sequences of a yeast plasmid and are distinguished by autonomous replication at high rates. Vectors containing an origin of autonomous replication transform yeasts at high efficiencies and are present in high copy numbers in the cell. Transformants with these replicating vectors are only stable on selective medium. Expression of the target gene and secretion of the primary gene product can be directed by including yeast transcription initiati-

on/termination and secretion signalling sequences into the vector. The food techno-
logist may require many different, important characters of an industrially used
yeast (Table 8.2). Recent reviews on the expression of foreign genes in yeast focus
on
- implementation of glycolytic and proteolytic activities,
- increased productivities of secondary gene products, and
- quality and aroma related compounds (Romanos et al., 1992; Mitter et al.,
  1993).

Some examples shall illustrate the latter topic: A brewer's yeast was complemen-
ted with the α-acetolactate decarboxylase gene (cf. Sect. 8.2.2; Fig. 2.2) and used
for main fermentation (Kronlof and Linko, 1992). The gene was successfully expressed and, thus, the formation of the undesirable diacetyl was lowered. The
lagering period (second fermentation), usually required for beer 'maturation', was
reduced from the conventional three to six weeks to about two to six days. No
adverse effects of the transformant on flavor formation and brewing performance
was found. Genetically modified strains of *S. cerevisiae* were compared with their
parent strains, and differences in 'flavor and aroma' between experimental wines
were found in sensory tests (Wightman et al., 1992). A gene of the L-leucine
pathway was introduced into sake yeast to overproduce methylbutanol, the immediate precursor of the fruity methylbutyl acetate (Hirata et al., 1992). Indeed, the
multicopy integrant was more productive and genetically stable. In a similar approach a yeast gene encoding the 3-desoxy-D-arabino heptulosonate-7-phosphate
synthase was transferred resulting in an overproduction of 2-phenyl ethanol (and
tyrosine) in sake (Fukuda et al., 1992). The mutated yeast gene was released from
feedback inhibition and was shown to serve as a dominant metabolic selection
marker. A wine yeast obtained additional endoglucanolytic properties by transfer
of a β-1,4-endoglucanase from a filamentous fungus (Perez-Gonzalez et al., 1993).
The expression of the gene and secretion of the enzyme produced a wine with
increased fruity aroma, very probably due to the liberation of odorous aglyca from
glycosides (cf.Sect. 7.2.1).

 To expand the range of carbon sources utilized *S. cerevisiae* strains were genetically engineered to express the *E. coli* β-galactosidase gene (Compagno et al.,

Table 8.2. Characters affecting the technological usefulness of a yeast strain

- Ethanol formation and tolerance
- Cell proliferation, growth at lower temperature
- Foaming during different phases of the bioprocess
- Aggregation and technical separability
- Metabolism of volatiles
- Metabolism of sulfur compounds
- Metabolism of acids compounds

1993), or the α-amylase gene from rice (Kumagai et al., 1993). Not only are the metabolic products of interest as precursors of volatile flavors, but also permit the recombinant yeast to use less expensive substrates, such as whey or lactose. Replacing antibiotic markers by metabolic markers (Inoue et al., 1993), shuttle vectors (Puta and Wambutt, 1992), and stabilization of recombinant DNA under bioprocessing conditions (Ramakrishnan and Hartley, 1993) will be given increased attention to establish industrial scale processes. Various strategies have been explored to stabilize yeasts during larger scale cultivations, among them reconstructed plasmids, differences in growth rates, and fed-batch culture (Kuriyama et al., 1992; Patkar and Seo, 1992; Hardjito et al., 1993).

So far, not much work has been devoted to recombinant yeasts distinguished by particular aroma formation capabilities. This must be blamed, to some extent, on the uncertainty about the impact of yeast on fermentation flavors in general (cf. Sect. 2.3). The above examples clearly indicate that tools exist for introducing predetermined genetic alterations into *S. cerevisiae*.

8.2.4 Higher Fungi

The tremendous potential of higher fungi for synthesizing aroma chemicals has already been discussed (cf. Sects. 5.2, 5.4). Genetic improvements are currently being impeded by the impossibility of breeding imperfect forms, the lack of metabolic information, and the problems associated with stable gene expression. Food grade fungi, such as *Aspergillus niger* ssp. *awamori* (Ward et al., 1993), and some penicillia have been selected for model experiments. Unexpected results, for example the loss of integration vector sequences, were reported. An intragenetic transfer of a glucoamylase gene into *A. oryzae* improved the fruity flavor of an experimental sake (Shibuya et al., 1992). The cDNA of a thaumatin gene (cf. Sect. 8.2.2) has been cloned and expressed in *A. oryzae* (Hahm and Batt, 1990). The fungal host secreted a sweet, immunoreactive protein. Since *A. oryzae* ist a food grade organism, it is ideally suited as a host for this plant gene.

8.2.5 Non-Food Species

Work with recombinant non-food species can indirectly contribute to our knowledge on the genetics of volatile odors and off-odors. Genes from *Lactobacillus helveticus* were expressed and studied in *E. coli* (Nowakowski et al., 1993). Other recombinant strains of *E. coli* possessed flavor aspects by accumulating L-phenylalanine (Dell and Frost, 1993), ethanol (Barbosa et al., 1992), or octanoic acid (Favrebulle et al., 1993). Increased production of ethanol by a recombinant *Klebsiella oxytoca* and hydroxylation of aliphatics or methylsubstituents of aromatics indicate the industrial direction of thrust (Wood and Ingram, 1992; Bosetti et al., 1992; Lonza, 1992). Oxyfunctionalizations are of particular interest to the flavor chemist (cf. Sect. 3.3; chapter 6). Recent work on streptomycete strains has indicated that not only the lactococcal formation of diacetyl, but also the genesis of geosmin (cf. Sect. 5.2) is plasmid encoded (Ishibashi, 1992). Giant linear plasmids were detected by pulsed-field gel electrophoresis. Strains cured of plasmids lost odor and pigment formation.

Classical and rDNA techniques have been applied to improve the formation of volatile flavors by starter cultures, to produce flavor related enzymes, and to generate flavor molecules as such. However, opportunities and larger scale applications are obviously out of proportion. Development costs and the persisting legal uncertainty (cf. Sect. 3.2) may be named as reasons. The most crucial factor appears to be the willingness of the average consumer to accept unbiased information on the safety of the novel foods (Teuber, 1993). Aroma compounds not only signal freshness and impart odor impressions, but may also possess antimicrobial or other physiological functions (cf. Sect. 1.5). This package of more or less consciously perceived properties has made aromas into the most intensively discussed group of food constituents. It will be up to future efforts to demystify the techniques of genetic engineering in order to overcome the surmised contradiction of biotechnological and hedonic advances.

9 Plant Catalysts

Most of the volatile flavors currently processed by the food industry are directly or indirectly based on the metabolic potential of plants (cf. Sect. 3.1). Various options exist for the deliberate use of this potential by the biotechnologist:

– Crude or partially purified enzymes, if unavailable from a microbial source, can perform transformations of selected substrates according to chapter 6.
– Incubated homogenates bring together previously compartmentalized substrates and enzymes for catabolic reactions.
– Intact parts of plant catalyze anabolic reactions from exogenous substrates.
– In vitro plant cells produce volatiles de novo or by transformation/conversion reactions.
– Genetically altered cells with improved flavor formation capability are multiplied by clonal micropropagation of known overproducing individuals, or obtained by selecting somaclonal variants from in vitro cultures, or obtained by direct transfer of heterologous genes.

In vitro plant cells are derived from a wounding induced tissue ('primary callus') by mechanical separation and continuous subculturing in a sterile environment. The nutrient media contain growth stimulants that favor morphological and biochemical dedifferentiation (cf. Table 4.2). Transfer to an agitated liquid medium and disintegration of the callus tissue finally reduces the macroorganism to the level of an eukaryotic single cell. Accordingly, each in vitro cell is provided with the complete genome and able to regenerate into a whole plant by organogenesis, or by somatic embryogenesis. This totipotency was translated into the idea that all metabolic functions of the parent plant could be expressed in the suspended cells at will. Three decades of cultivation of essential oil plants in vitro have, however, demonstrated that the induction of aroma pathways usually poses serious problems. Improvement of the cultivation conditions has sometimes led to low product concentrations; if a higher productivity has been achieved, rapid turnover and cytotoxicity of the target compound prevented accumulation; if accumulation was obtained by technical measures, the different spectrum of volatile products or the instable generation created new hurdles.

Incorporation of new DNA into the plant cell has been achieved directly, for example by using ballistic microprojectiles coated with nucleic acids or by microinjection (into protoplasts or intact cells), or by the use of *Agrobacterium* vectors. Previous work was directed at implementing genes that give resistance to viral or insect attack or to herbicides. *Agrobacterium rhizogenes* can transform *So-*

lanaceae (and other families) into rapidly growing and productive 'hairy root' cultures. Secondary metabolites typically produced in plant roots, such as nicotine and other alkaloids were accumulated by these *vitro* organs. Unfortunately, most volatile flavors originate in the aerial parts of plants.

9.1 Plant Enzymes

According to the 'flavorese enzyme' concept, crude plant enzymes, when added to a thermally processed commodity, act on nonvolatile, stable flavor precursors, converting them into volatile flavors. Thus, the lost part of the original flavor is regenerated (Hewitt et al., 1956; Weurman, 1961). The chemical classes of neither the enzymes prepared nor the substrates transformed were investigated in the early reports. The patent of Motzel and Baur (Procter and Gamble, 1965) described products normally discarded as waste in the preparation of preserved fruits as an economic source of enzymes. Exogenous substrates were used so that metabolic relationships led to volatile flavors: Additions of fatty acids, amino acids, and members of the citric acid cycle resulted in different sensory preferences in flavor regeneration studies. Attempts to improve the aroma of processed plant materials covered apple, apricot, banana, blackberry, blueberry, cherry, citrus fruits, peach, pear, pineapple, plum, raspberry, strawberry, and vegetables, such as carrot and tomato (Askar and Bielig, 1976; Crouzet, 1977). For example, enzymes were prepared as an acetone powder from the cone of raspberry, and released volatiles upon incubation with unripe fruits or fruit pulp (Bruchmann and Kolb, 1973). A number of enzymes were classified in the crude preparations, but, in the light of recent results, it appears likely that much of the reported aroma inprovements rested on the hydrolysis of glycosidic precursors (cf. Sect. 7.2.1).

The crucial problem of the flavorese enzyme concept is still the impossibility of isolating and operating an active multienzyme complex stably in vitro. Even simple, aroma relevant metabolic steps like the transformation of pyruvate into acetyl-CoA (and CO_2) require several enzymatic activities and the associated cofactors. As a result, only one pathway is currently further researched for commercial applications: The lipoxygenase/hydroperoxide lyase (or dehydrase) mediated catabolism of C_{18}-unsaturated fatty acids.

9.1.1 Lipoxygenase-Derived Aromas

The typical fresh and pleasant top notes of crushed fruits (and many green parts of plants, 'green' note) arise from volatile C_6-compounds, such as (E2)-hexenal ('leaf aldehyde'), and (Z3)-hexenol ('leaf alcohol') (cf. Sects. 1.5, 5.3.2). Disintegration of the plant cell by processing enables an immediate contact of the thylakoid bound enzyme system and its precursors. Exogenous and intercellular oxygen act as a cosubstrate in the hydroperoxidation step. Meanwhile, many biogenetic details have been elucidated (Hatanaka, 1993), but enzyme activity levels in the usual agricultural waste streams have permitted only moderate yields. The lipoxygenase isoenzymes of soybean flour have been studied, because they are particularly stable and contribute undesirable beany off-notes to soya milk and soybean supplemented bread (cf. Sect. 5.5). Differences existed among the isoenzymes in terms of generation of volatile flavor and aroma (Addo et al., 1993). Higher levels of alcohols were found in soy flour amended bread dough than in soy flour homogenates alone. The genetic elimination of the LOX 2 and LOX 3 isoenzymes made LOX 1 more effective in regard to volatiles production (Pereira et al., 1992). Another group came to the opposite conclusion using a different, LOX 2 and LOX 3 deficient cultivar (Kitamura et al., 1992).

Apple pomace, a waste stream of juice manufacturing, has been recognized as another rich source of lipoxygenase derived volatiles. French, German, Hungarian and Japanese groups have tried to develop industrial scale processes by identifying the rate limiting factors of this reaction. Only recently, counteracting the oxidation of phenolic compounds by adding sulfur dioxide and ascorbic acid was shown to accelerate the bioconversion (Almosnino and Bélin, 1991). In a comparative study, a non-specified apple cultivar yielded only medium amounts of n-hexanal (Muller and Gautier, 1994). Fatty acid hydroperoxides were prepared separately (using soya lipoxygenase) and added to various plant homogenates. Alfalfa leaves (1360 mg n-hexanal$\times$kg^{-1}) and green peppers were identified as tissues bearing the highest hydroperoxide lyase activity. When linolenic acid C_{13}-hydroperoxide was used, the levels of (E2)-hexenal were lower, although efficient producers of n-hexanal also efficiently converted the threefold unsaturated substrate. The corresponding C_6-alcohols were obtained using the dehydrogenase activity of fermenting baker's yeast according to the examples given in Sect. 6.3 (Pernod-Ricard, 1991). (Z3)-hexenol prepared by this method was compared with mint derived and synthetic samples using ^{2}H-NMR (Muller and Gautier, 1993). A strong depletion was found for the deuterium in position 1 in the yeast product, and in positions 3 and 4 in the chemosynthesized counterpart. Isotope effects related to the different synthetic pathways were discussed.

The volatile compounds characterizing the aroma of fresh fish result from lipoxygenase induced reactions rather than from mere autoxidation, as reviewed by Josephson and Lindsay (1986). In a rapid catabolism of polyunsaturated fatty acids volatiles are formed in gill homogenates via 12- and 15-hydroperoxides (German et al., 1992). By analogy to the pathways proposed in wounded plants and mushrooms (cf. Fig. 5.6, Fig. 9.1), at least four different enzymes are required to explain the full range of volatile breakdown products. Compounds, such as trimethylamine or other amines, indicative of microbial spoilage, and hepta- or decaenols, indicative of autoxidative lipid degradation, were not found in samples of fresh fish (German et al., 1991). GC-Olfactometry (cf. Sect. 1.2) showed the major constituents of the extracts to be contributory to fresh fish flavor in their actual concentrations. It was some trace compounds that exerted even stronger odor impressions: (E2)-pentenal (green apple), (E2,Z6)-nonadienal (cucumber), 1-octen-3-one (field mushroom), (1,Z5)-octadien-3-one (geranium leaves), (1,E3,Z5)-

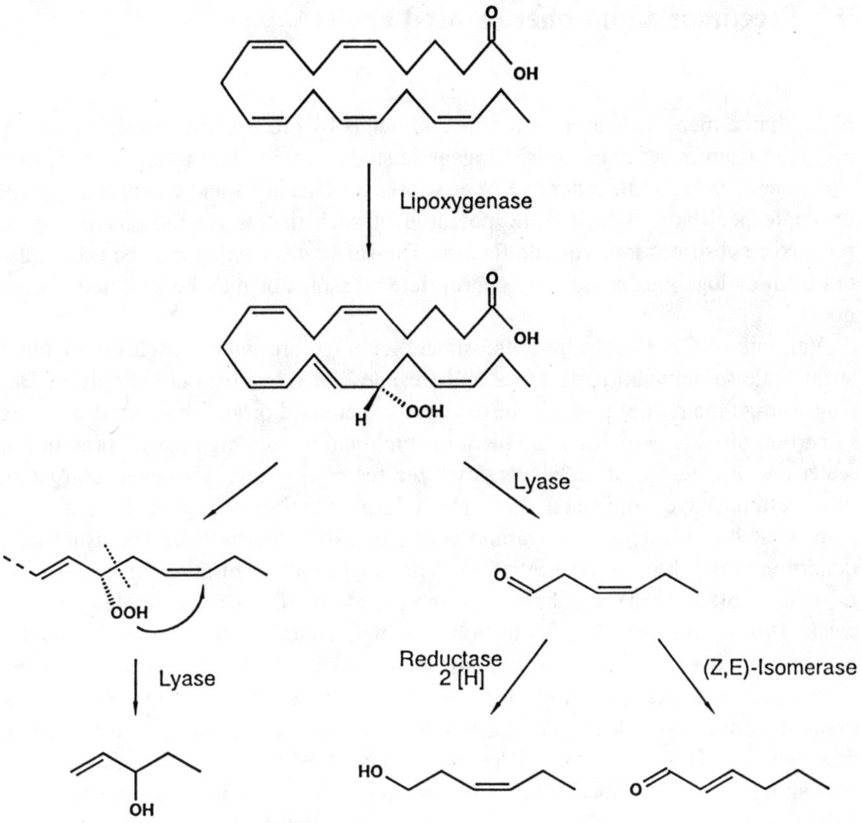

Fig. 9.1. Formal derivation of five and six carbon volatiles of fresh seafood (German et al., 1992)

undecatriene (balsamic), and (1,E3,Z5,Z8)-undecatetraene (seaweed). The concentrations of these volatiles in gill homogenate were up to 200fold higher than the respective values of whole fish. The strong cucumber odor of various species of smelt *(Osmeridae, Retropinnidae)* was identified as (E2,E6)-nonadienal by independent groups (Zhang et al., 1992; McDowall et al., 1993). Waste materials of fish processing could thus be developed into biocatalysts for the generation of seafood volatiles by making use of the only non-plant enzyme system yielding volatile impact components. The application of lipoxygenases for the degradation of phenyl propanoids to phenyl aldehydes was described in Sect. 5.4.1.

9.2 Precursor Atmosphere Stored Fruits

More than a mere formation of aroma compunds by precursor degradation can by expected from intact plant cells. Biogenetic studies using plant tissue preparations (Schreier, 1984) and the attentive observations of Guadagni and coworkers (1971) on apple peel have indicated the potential of such tissues for the conversion of precursor substrates into volatile flavors. The substrates needed may be accumulated from endogenous sources by appropriate measures or may be supplied exogenously.

Variants of CA ('controlled atmosphere')-storage are widely used on an industrial scale to maintain appearance and freshness of many fruit commodities. During storage the partial pressure of oxygen is decreased or/and the partial pressure of carbon dioxide is increased, which, in combination with high partial pressure of water and low temperature, slows down the respiration rate. However, consistent observations have confirmed that aroma formation from fatty acids and, to a somewhat lesser extent, from amino acids, is also suppressed by the usual CA-conditions (Brackmann et al., 1993). The oxygen concentration of storage atmospheres effected the post-harvest formation of volatile esters in apples (Hansen et al., 1992). The volatile composition of nonclimacteric fruit depended similarly on the composition of the storage atmosphere (Shaw et al., 1992a). Low oxygen partial pressure increased the ester content of oranges (Shaw et al., 1992b). Increased carbon dioxide concentrations were favorable for the aroma of stored strawberries (Table 9.1), but under conditions of MA ('modified atmosphere')-packaging no improvement of the flavor quality was possible (Shamaila et al., 1992). As flavor accumulation in fruits is closely related to senescence, the usual storage and packaging strategies aiming at a prevention of maturation cannot be expected to increase aroma scores (Ueda et al., 1993).

Table 9.1. Concentration of impact volatiles of strawberries cv. *Red Gauntlet*

Compound (μg 100 g^{-1})	Untreated	plus CO_2
Ethyl hexanoate	20	59
Hexyl acetate	1	12
Mesifurane [a]	81	918
Furaneol[b]	48	226
Methyl (E)-cinnamate	70	146
4-Decanolide	266	1017
5-Decanolide	6	26

[a] 2,5-dimethyl-4-methoxy-3(2H)-furan-3-one and the
[b] 4-hydroxy analogue

Hypoxia of climacteric fruits was traditionally related to the formation of ethanol and acetaldehyde (Nanos et al., 1992). Only recently, a comparative study demonstrated that tissue of apple fruit responded to anaerobic conditions in a way similar to rice roots: Lactic fermentation was induced immediately, whereas alcoholic fermentation was much less pronounced (Andreev and Vartapetian, 1992). There was no single universal mechanism of induction of the nonrespiratory pathways in organs of higher plants upon transfer to oxygen deficiency conditions. Other mechanisms involved in low oxygen metabolism are currently under investigation (Kennedy et al., 1992). As a preliminary result, it appears difficult to mobilize aroma precursors under established conditions of CA or MA storage.

Fruits are fumigated with low-molecular weight n-alkanals to prevent fungal decay (Mattheis and Roberts, 1993). Concurrent change of the aroma profile as a result of such treatments was observed. In ripening strawberry a 90% drop in the concentration of L-alanine during the phase of active ester formation was interpreted as a depletion of an endogenous precursor pool (Perez et al., 1992). This led, together with the above mentioned anabolic properties of fruit tissues, to develop a biotechnological concept that used intact fruits as an immobilized plant catalyst with an active aroma metabolism (Berger et al., 1992). Called PA ('precursor atmosphere') – storage, the process included exposing fruits to a modified atmosphere containing vapors of volatile aroma precursors. Thereby, the problems of microbial interference and larger scale production associated with defined tissue disks of fruits were eliminated. Increasing the substrate concentration immediately increased the concentration of the respective carboxylic esters in apple; for example, the concentration of ethyl 2-methylbutanate, an impact of many apple cultivars, increased by a factor of more than 50 in a storage environment enriched with ethanol. Different precursors or mixtures of precursors can be applied to shift ester formation into the desired direction (Table 9.2). Compared to thermal or enzymatic formation of carboxylic acid esters (cf. Sect. 7.1.3), PA-storage creates a balanced mixture of esters by making use of endogenous (chiral) and exogenous substrate pools.

The respiratory increase of climacteric fruit is seen as a part of the efforts of a senescent cell to maintain homeostasis. The de novo formation of volatiles during

Table 9.2. Concentration of volatiles of apples cv. *Jonathan* (PA-storage 3d, 10 mmol n-butanol kg^{-1}, 20 °C)

Ester (μg 100 g^{-1})	Untreated	PA-storage
Ethyl butanoate	69	460
Butyl butanoate	27	2280
Butyl 2-methyl butanoate	33	434
Butyl hexanoate	185	963
Hexyl butanoate	108	<1

PA-storage can be integrated into this view, because the excessive degradation of membrane polymers and storage compounds implies the need to get rid of osmotically and toxicologically active constituents, such as free fatty acids. One means to metabolic relief is volatilization and removal from the cell using the steep, natural concentration gradient. Hence, the headspace above PA-stored fruits contains large amounts of volatiles. Recovery from the gas phase can be achieved using fluidized bed columns filled with polymer adsorbents. The major portion of volatiles is retained in the fruit and can be transferred into processed, aroma-rich products such as jam or juice.

PA-storage has been shown to enhance the ester content of apples, pears, cherries, strawberries and bananas. When ripe bananas were sliced and freeze-dried, a substantial loss of numerous volatiles was recorded (Berger et al., 1986). Banana fruits, pretreated by PA-storage and processed to a freeze-dried product like the untreated control, showed residual levels of methylbutyl esters that approximated those of the fresh fruit. Nonclimacteric fruits, such as orange, are not excluded from PA-storage stimulated flavor changes (Lizotte and Shaw, 1992). Strawberry fruits contained elevated levels of ethyl esters upon exposure to ethanol, and various other impact volatiles showed also increased concentrations (Table 9.3). An active alcohol acyltransferase has been isolated from the fruit and characterized as to substrate specificity and physical properties (Perez et al., 1993), but obviously different pathways were simultaneously stimulated in the ethanol incubation experiment. As a result, managing of endogenous enzyme chains and their substrates compensated for aroma deficiencies. The origins of this lack of flavor are irrelevant and could be premature harvesting, improper handling, or physical losses during thermal operations.

Table 9.3. Concentration of impact volatiles of strawberries cv. *Senga sengana* (PA-storage 24 h, 8 mmol ethanol kg^{-1}, 20 °C)

Compound (µg 100 g^{-1})	Untreated	PA-storage
Ethyl butanoate	390	885
Ethyl (E)-cinnamate	102	173
Mesifurane[a]	75	87
Linalool	23	46
4-Decanolid	67	110

[a] cf. Table 9.1

9.3 In Vitro Plant Cell Culture

Plant cell culture refers to the cultivation of single cells, tissues and whole organs and is distinguished from plant enzyme or fruit based bioprocesses mainly by sterile working conditions and regeneration of the catalyst. The advantage of continuous growth and process operation is opposed by the often very low levels of expression of secondary pathways. Suspended in vitro plant cells have lost their previous position and interconnection and are forced to divide continuously by growth regulators incorporated into the nutrient medium. During subculturing a heterogeneous population develops that provides the researcher with the same potential that mutants have conferred on microbial biotechnology. All the different genotypes, however, appear very homogeneous by the pronounced primary metabolism that ensures survival in the synthetic environment. Therefore, any aroma application has to experimentally overcome fundamental metabolic obstacles.

9.3.1 De Novo Syntheses

A remarkable number of publications has described aliphatic, aromatic, phenylpropanoid, oligoprenoid and S-containing volatiles in plant cell cultures. Recent examples have been compiled in Table 9.4. Not only seed plants, but also liverworts were reported to form odorous volatiles (Ono et al., 1992). Nonvolatile flavors, such as neohesperidin and naringin, major flavonoids of bitter orange were also found in callus cells (Del Rio et al., 1992). A closer look at the current literature, however, shows that only few potentially viable cellular systems exist, and

Table 9.4. Volatiles from plant cell culture

Species	Reference
Elettaria cardamomum	Bajaj et al., 1993
Allium cepa	Collin and Britton, 1993
Allium cepa	Gbolade and Lockwood, 1992
Melissa officinalis, Petroselinum cripum	Gbolade and Lockwood, 1991
Taraxacum officinale	Hook et al, 1991
Vanilla planifolia, Mentha sp.	Knorr et al., 1993
Pogostemon cablin	Nabeta et al., 1993
Iris sibirica	Para and Baratti, 1992
Allium sp.	Prince and Shuler, 1993
Melissa officinalis	Schultze et al., 1993
Thymus vulgaris	Tamura et al., 1993

commercialization is not certain even for compounds, such as capsaicin and vanillin that have received worldwide acclaim. Cultured cells frequently displayed lower levels and a more restricted range of volatile flavors than the parent plant. Onion cells, for example, produced typical volatiles but lacked the lacrimatory factor (Gbolade and Lockwood, 1992). Low equilibrium concentrations were due to high catabolic activities in lemon cells (Berger et al., 1991). Some authors have successfully induced secondary metabolite formation by starting a partial differentiation during immobilization. Though the immobilized state complies better with the normal status of a plant cell, enhanced flavor formation has not regularly been found. Onion cells have been researched more frequently, because much is already known about the chemistry of the nonvolatile flavor precursors (Fig. 9.2). Feeding of amino acid precursors enhanced the concentration of the sulfoxides and altered the flavor spectrum. In the presence of L-cystein the amount of the propyl derivative increased tenfold, whereas L-methionine not only failed to stimulate, but actually decreased the concentration of this derivative.

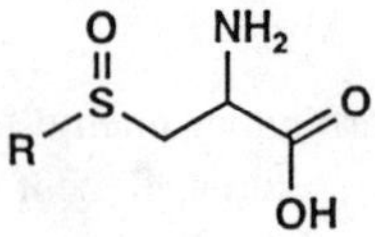

Fig. 9.2. (+)-*S*-alkyl cystein sulfoxides produced by in vitro rootlet cultures of *Allium* (R=methyl, propyl, propenyl, allyl) (Prince and Shuler, 1993)

9.3.2 Differentiated Cells

Cells within a particular plant organ perform specific functions that may not be common to all parts of the organism. This raises the question of how to attain a favorable status of cytodifferentiation in vitro.

Efforts were made to induce morphological differentiation by manipulation of the phytoeffector concentrations and ratios. Biochemical differentiation, often at the expense of the loss of an activly growing cell culture, was the result hoped for. A strong auxin induced the formation of glandular trichomes on callus cells of black currant (Joersbo et al., 1992). The number of these oil gland equivalents decreased upon addition of cytokinins. The gland forming capability faded during consecutive subculturing, an observation that can be generalized for many differentiation processes. Stable proliferation of shoots and nutritional conditions favoring shoot elongation resulted in a significant increase of monoterpene formation in *Lavendula latifolia* (Calvo and Sanchez-Gras, 1993), and in *Zingiber officinale* (Ilahi and Jabeen, 1992). Similar patterns of essential oil constituents were found in the differentiated callus or differentiated organ and the parent plant. In undifferentiated cultured cells of *Mentha* sp. the amounts of volatiles produced were often less than one 100th of those found in the mother plant. The synthetic potential was regenerated in adventitious shoots and regenerated plantlets of *Mentha arvensis*

(Kawabe et al., 1993). The ratio of the predominant monoterpenes changed with the concentration of 6-benzylaminopurine, a cytokinine. As L-menthol represents one of the bulk chemicals of the aroma industry (cf. Sect. 7.1.2), there is the possibility that advanced techniques for culturing differentiated cells of essential oil plants may be used for the generation of precious essential oils by an in-vitro system. Shoot and root forming callus tissues of onion converted S-allyl-L-cystein to alliin, the corresponding sulfoxide (Ohsumi et al., 1993; Fig. 9.2). Conversion was quantitative and yielded pure (+)-stereoisomer alliin.

Aerial roots of vanilla transformed ferulic acid into vanillin (Westcott et al., 1994; Fig. 5.10). A charcoal trap was used as a product reservoir, and a product concentration of about 40% of that present in mature vanilla beans could be obtained 5 to 10 times faster than its normal synthesis in the beans. As for flavorese enzymes, the activity of the biocatalyst gradually declined with reuse, and the wheel of ideas comes full circle (cf. Sects. 5.4.1, 9.1).

9.3.3 Conversion and Conjugation of Substrates

In analogy to microbial conversions (cf. chapter 6), but with superior biochemical potential, cultured plant cells can be employed to conduct bioconversions (Table 9.5). Again, low predictability of the course of the reaction and the chance to encounter novel catalytic directions go hand in hand. Immobilized cells of carrot enantioselectively transformed aromatic ketones, such as acetophenone or propiophenone, into the corresponding (S)-alcohols, as yeasts do. The same substrates were transformed to (R)-alcohols in the presence of immobilized *Gardenia jasminoides* cells (Akakabe and Naoshima, 1993). This anti-Prelog (cf. Fig. 6.11) behaviour was explained by a two step reaction. Initial reduction produced racemic 2-phenalkyl alcohol, followed by a stereoselective oxidation of the (S)-enantiomer to the ketone. Microbial dehydrogenases that violate Prelog's rule have become known too. Immobilized plant cells offer opportunities to improve

Table 9.5. Bioconversions of aroma chemicals by cultured plant cells (including algae)

Substrate(s)	Species	Reference
Acetophenone	*Gardenia jasminoides*	Akakabe and Naoshima, 1993
Aromatic aldehydes	*Dunaliella tertiolecta*	Noma et al., 1992
Carveols	*Euglena gracilis*	Noma and Asakawa, 1992
Citronellol derivatives	*Spirodela punctata*	Pawlowicz et al., 1992
Germacrone	*Curcuma zedoaria*	Sakui et al., 1992
Carvone, pulegone	*Nicotiana tabacum*	Suga and Tang, 1993
Menthyl acetates	*Mentha* sp.	Werrmann and Knorr, 1993

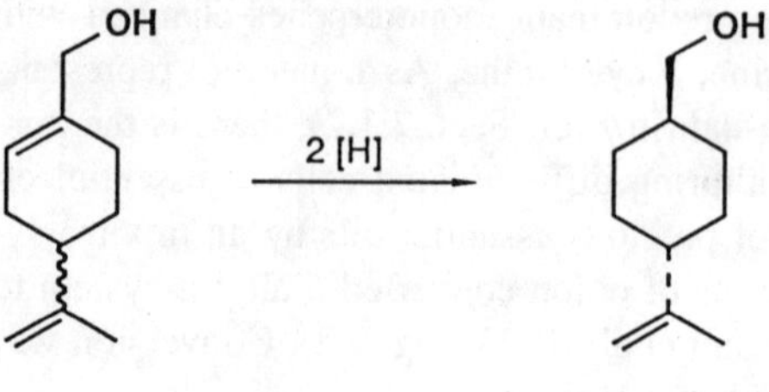

Fig. 9.3. Enantiofacial preference of *Euglena* cells in the transformation of perilla alcohol to *trans*-shisool (Noma et al., 1992)

the reaction specificity (Suga and Tang, 1993): By shifting the pH of the culture medium near to the pH optimum of an enone reductase, the removal of the carbon-carbon double bond of (-)-carvone and (+)-pulegone by tobacco cells was turned into the predominant reaction. Unicellular microalgae, such as *Dunaliella* and *Euglena* (Table 9.5), have also been demonstrated to transform monoterpenes diastereo– and enantioselectively (Fig. 9.3). In the study of Noma and Asakawa (1992) some carveol isomers were not transformed at all, while others resulted in a series of volatile products. A technical scale process will have to solve the problem of illumination of these photosynthetic organisms.

Cyclic monoterpenols or terpenols devoid of α, β-unsaturation better resist an enzymatic oxidation and are channelled into another detoxification pathway: gly-cosilation (Table 9.6). The difficulties of a specific chemosynthetic glycosilation and the technical role of glycosides as an aroma depot have been mentioned above (cf. Sect. 7.2). Similar considerations apply to glycosyl esters of multifunctional acyl moieties. As the mobilization and utilization of numerous plant metabolites depend on these enzymatic reactions, excellent reaction rates and specificities have been reported. 85% of the vanillin supplemented were converted into the glucoside by coffee cells within one day (Kometani et al., 1993). The β-glucoside retained the antimutagenic and antimicrobial activities of the aglycon. β-glucosides were found most frequently, but other mono- and diglycosyl residues have also been identified (Orihara and Furuya, 1993). If a given metabolite lacked the required functionality, plant cells were observed to hydroxylate and then glycosylate the

Table 9.6. Glycosilations performed by cultured plant cells

Subsate	Specics	Reference
Glycyrrhetinic acid	*Glycyrrhiza glabra*	Hayashi et al., 1992
Butanoic acid	*Nicotiana plumbaginifolia*	Kamel et al., 1992
Vanillin	*Coffea arabica*	Kometani et al., 1993
Borneol	*Eucalyptus perriniana*	Orihara and Furuya, 1993

compound at the newly formed hydroxy function (Hayashi et al., 1992). Glucuronylation of terpenoids by suspended plant cells was reported by the same autors.

In vitro plant cells offer numerous advantages for biosynthetic studies that are indispensable for preparing biotechnological applications. The conversion of labeled mevalonate to deuterated sesquiterpenes by *Perilla frutescens* (Nabeta et al., 1993), and the complex transformation steps from a monocyclic sesquiterpene precursor to other sesquiterpene classes (Sakui et al., 1992) have confirmed hypothetical hydride shift and transannular cyclization reactions, although the results should be transferred to in vivo conditions with care. Deuterated metabolites were evaluated using coupled gas chromatography – mass spectrometry (Arigoni et al., 1993). Results indicated that cyclization of geranyl diphosphate proceeded according to the currently accepted mechanism. Contradictory, previously obtained data using natural deuterium abundance NMR spectroscopy await further clarification. Very few experiments have been conducted on an enzymatic level. A camphor hydroxylase from sage was accumulated in cell culture, and the enzyme was characterized physically, chemically, and immunologically (Funk and Croteau, 1993). The inducable enzyme had the properties of a microsomal cytochrome P-450 monooxygenase and was believed to share common features with microbial and other plant monoterpene hydroxylases. More biochemical information of this type will be required to reach the ultimate goal of plant cell culture technology: the transfer of plant genes into easier to handle, lower organisms. Meanwhile, many attempts are made to increase the level of gene expression and accumulation of target products by empirical means.

9.3.4 Key Parameters for the Production of Aromas

The tremendous diversity and industrial importance of plant volatiles, unequalled by any other biosystem, and the recurrent confirmation of the genetic totipotency of in vitro cells through regeneration experiments (cf. Sect. 9.3.2) have maintained the interest in plant cell culture. Flavor formation may be expressed at low levels, for limited periods, or in quantitative ratios differing from the whole plant. Many authors have reported on the 'optimization' of secondary metabolite production, but, on a closer look, only few species, such as *Lithospermum erythrorhizon*, were submitted to systematic and comprehensive procedures to evaluate improved in vitro conditions (Buitelaar and Tramper, 1992). Without an integrative view on the isolated in vitro cell, misleading interpretations are bound to discredit the technique as such. For example, the problem of intracellular storage of target compound was claimed to be solved by efficient methods of membrane permeabilization ignoring adverse effects on cell vitality by the partially irreversible character of the treatments. The most promising strategy to increase yields appears to be a combination of stimulating certain pathways by the creation of stress conditions together with rapid product removal from the producer cell (Brodelius and Pedersen, 1993).

Lipophilic liquids have been used as a means to prevent substrate or product inhibition or degradation. Little has become known on the biocompatibility of the second phase. The solvents chosen for supplementing lemon cells with β-carotene all affected the vitality of stationary phase cells after a short exposure (Berger, 1994). Methanol and dimethyl formamide were the least critical solvents, but neither partition coefficient nor dielectric constant data could be correlated with the cytotoxicity of a series of solvents. Bioconversions of terpenols by in vitro apple cells, by contrast, were retarded by low concentrations of methanol, while dimethyl formamide, dioxane, and to a lesser extent ethanol were tolerated in concentrations up to 1% v/v. The terpenol substrates were harmful in low mmolar concentrations. Dimethyl sulfoxide, a solvent with permeabilizing properties, reduced the activity of total exo-peroxidase, used as a metabolic marker, even in concentrations of 0,1% v/v.

The beneficial effects on product yields of other synthetic accumulation sites, such as charcoal, triacylglycerols, or RP materials may be interpreted in terms of both product protection and decreased evaporation losses. Synthetic triglycerides can be easily administered, pumped and distilled, but may not be metabolically inert (Berger, 1994; Table 9.7). A metabolic overflow upon addition of excess carbon source evidently resulted in a vigorous production of volatiles. Even volatile metabolites untypical of the whole plant can occur after the homeostatic equilibria have been brought out of balance. Solid lipophilic polymers have been introduced into plant biotechnology, though they are often loaded with cytotoxic contaminants, pose separation problems, and show unsatisfactory recovery. From a technical point of view switching a fixed bed of polymer into an external loop of a bioreactor appears promising (cf. Sect. 10.3). Aqueous two-phase polymer systems comprising polyethylene glycol and dextran have also been described.

The conventional approach to improve the net production rates of target compounds has been to modify the chemical environment of the cell. Time consuming

Table 9.7. In vitro cultured cells of apple use exogenous triglyceride as a substrate and synthetic site of accumulation of volatile flavor (Berger, 1994)

Compound	$[\mu g\ L^{-1}]$
Methyl *n*-octanoate	171
Ethyl *n*-octanoate	81
Methyl *n*-decanoate	92
Ethyl *n*-decanoate	276
n-Decanol	214
n-Nonanol	59
n-Decanol	22
4-Hexanolide	240
4-Octanolide	70

Table 9.8. High-yielding plant cell cultures

Compound	Species	Yield [g L^{-1}]	Reference
Berberine	*Coptis japonica*	7	Fujita and Tabata, 1987
Rosmarinic acid	*Coleus blumei*	5.6	Kesselring, 1985
Shikonine	*Lithospermum erythrrorhizon*	3.5	Tabata and Fujita, 1985
Jatrorrhizine	*Berberis stolonifera*	3.3	Stöckigt and Schübel, 1989

and somewhat random, these efforts have succeeded in establishing some highly productive bioprocesses (Table 9.8). Topics under discussion were inexpensive carbon sources, buffered media, gelling agents, conditioning factors, and phytoeffectors. Representatives of the latter class of compounds control differentiation (cf. Sect. 9.3.2), are usually essential for growth, and are known to effect metabolite levels in nondifferentiated cells. One should be aware that exogenous phytoeffectors just add to the levels of endogenous regulator compounds. As a rough rule, excellent growth rates exclude secondary metabolite formation in plant cell culture (cf. Sect. 5.6). It is a recent finding that certain plant volatiles, particularly mono- and sesquiterpenes, are formed in narrow windows of phytoeffector concentrations (Reil, 1993). Only 3 out of 36 different phytoeffector ratios examined permitted the formation of limonene in callus cells of rosemary (Table 9.9). Production of volatiles and letal conditions were often close, again indicating the extreme metabolic situation required for the generation of plant aromas.

Stress on cultivated plant cells (by definition) restricts growth and may elevate the level of expression of inducable enzymes. Biotic factors (elicitors) and abiotic stress factors can, thus, stimulate the formation of typical or novel metabolites, among them volatiles. Various modes of elicitation can be distinguished:

- The elicitor molecule is a microbial metabolite and recognized by specific membrane receptors (Table 9.10).
- Microbial hydrolases release plant cell wall constituents that act as elicitation molecule (Table 9.11).
- Plant enzymes liberate microbial cell wall fragments that, in turn, induce phytoalexin formation (Table 9.12).
- The elicitor molecule is a plant metabolite generated in response to an exogenous stress factor (Table 9.13).

By now, the application of elicitors has developed into a key to success. Depending on the type of elicitor different secondary metabolites may accumulate. Common features of elicited plant cells were dosage dependent response, product

Table 9.9. Concentration of limonene [µg kg^{-1}] in vitro cells of *Rosmarinus officinalis* (SH medium; illumination 3000 lux)

Concentration of naphthyl acetic acid [mg L^{-1}]	0.05	0.1	0.25	5	0.75	1
Concentration of kinetin [mg L^{-1}]						
< 1	-	-	-	-	-[a]	-[a]
1.0	-	550	660	580	-	-[a]
> 1	-	-	-	-	-[a]	-[a]

[a] lethal conditions

Table 9.10. Volatiles produced by *Coleonéma album* callus cultures related to type and concentration of gellant (SH 4 medium; illumination 3000 lux)

Compound [µg kg^{-1}]	Agar Agar				Gellan Gum		
	3	6	12		3	6	12 [g L^{-1}]
Myrcene	45	79	134		850	1080	1800
α-Phellandrene	93	146	245		1500	1960	3420
Limonene	211	330	75		3600	4700	8400

formation within a few hours, and degradation of accumulated product after one to two days. Comparing an algal and a bacterial gelling agent a concentration dependent elicitation and the superior activity of the microbial elicitor were demonstrated (Table 9.10). Caused by the ecological background, compounds produced in the course of an elicitation process are often released into the surrounding medium (Table 9.11). It was frequently observed that elicited cells produced compounds that were thought to be typical of completely different taxa. For example, the phthalides identified in elicited parsley (Table 9.11, Fig. 9.4), are more abundant in celery bulbs, and vanillin, unlike other phenols, has never been identified in tomato (Tables 9.12, 9.13). An enhanced incorporation of vanillin and related compounds into the cell wall of *Lycopersicon peruvianum* was observed upon treatment of cells with autoclaved mycelium or culture filtrate of *Fusarium oxysporum*, a known tomato pathogen (Beimen et al., 1992). Not unexpectedly, chitosans of different origins had elicitor activity, with microbial chitosans being more effective (Knorr et al., 1990). Some elicitations occurred irrespective of the growth phase, while others depended on subcellular differentiation during aging (Tani et al., 1993). The considerable overlap of elicited volatiles and of bioactive plant volatiles (cf. Sect. 1.5) cannot be overlooked. The signalling roles of volatile plant emissions is now, supported by the upsurge of headspace purge and trap techniques and related methodology, more and more recognized (Loreto and Sharkey, 1993).

Table 9.11. Volatiles produced by suspended cells of *Petroselinum crispum* 24 h upon addition of a fungal homogenate (*Polyporus umbellatus*; SH4 medium; 3000 lux)

Compound [µg kg^{-1} or L^{-1}]	Intracellular	Extracellular
3*n*-Butylphthalide	80	420
Sedanolide	70	450
Sedanenolide	n.d.	170
(Z)-lingustilide	30	80
Elimicin	150	n.d.

Table 9.12. Volatiles produced by suspended plant cells in the presence of heat inactivated fungal mycelium (*Tyromyces* sp., 35 mg 100 ml^{-1})

Species	Induced product	Concentration [µg kg^{-1}]
Rosmarinus off.	3-Methyl-3-butenol	750
	3-Methyl-2-butenol	480
	Myrtenol	640
Lycopersicon esculentum	(E2)-nonenal	1350
	(E3)-nonenal	810
	(Z2, E6)-nonadienal	1060
	Vanillin	3450

Table 9.13. Volatiles produced during co-cultivation of suspended cells of tomato and sage (1 g 100 ml^{-1}]

Compound	Concentration [µg kg^{-1}]
Limonene	330
(Z2)-nonenal	2,640
(E2)-nonenal	800
(E3)-nonenal	2,440
(Z2, E6)-nonadienal	680
4-Vinylphenol	1160
Vanillin	1000

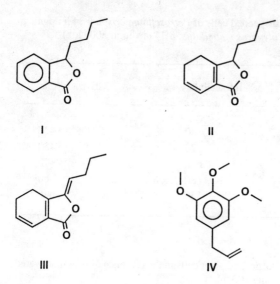

I

II

III

IV

Fig. 9.4. Volatiles in elicited suspension cultured cells of parsley (cf. Table 9.11). I=3-*n*-Butyl phthalide, II=Sedanenolide, III=(Z)-Lingustilide, IV=Elimicin

9.3.5 Phototrophic in Vitro Cells

Photoheterotrophic in vitro cells at high densities, the standard experimental system for many years, experience conditions resembling those existing in root cells. Many reports on successful secondary metabolite production accordingly refer to root-synthesized compounds (Table 9.8). The synthesis of volatiles, however, is usually associated with the green parts of the plant. Leaves of *Ginkgo biloba*, for example, accumulated terpenes with increased photonic level, and not related to leaf growth or a specific state of organ development (Flesch et al., 1992). The unique ability of a plant cell to synthesize complexest compounds using light and carbon dioxide founds the existence of higher life forms, but commercial applications of in vitro systems remained rare.

Light induced cytodifferentiation and accumulation of volatiles was observed in photomixotrophic cell cultures of *Coleonéma album*, a Rutaceae (Reil et al., 1994). The volatile oil from shoots was composed of monoterpenes, sesquiterpenes, and phenylpropanoids. A photomixotrophic cell line, established from leaves, contained a heterogeneous population of plastids and synthesized an essential oil that differed from the leaf oil of the intact plant (Fig 9.5). Substitution of 2,4-dichlorophenoxy acetic acid and *p*-chlorophenoxy acetic acid by naphthalene acetic acid and a high kinetin concentration favored the formation of essential oil (Fig. 9.6). The parallels to the behaviour of the in vitro cells of rosemary are evident (Table 9.9). The photoheterotrophic callus, when grown under otherwise identical conditions, was devoid of any differentiated plastids and of volatile con-

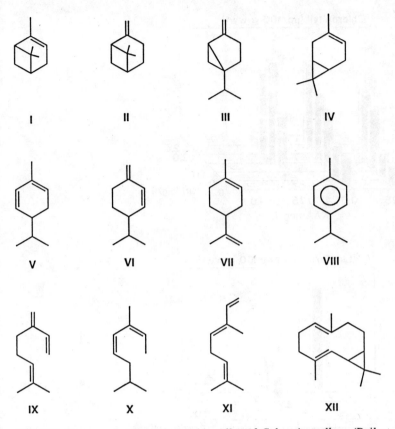

Fig. 9.5. Terpenes in photomixotrophic callus of *Coleonéma album* (Reil et al., 1994). I/II = α/β-Pinene, III = Sabinene, IV = δ-3-Carene, V/VI = α/β-Phellandrene, VII = Limonene, VIII = *p*-Cymene, IX = Myrcene, X/XI = α/β-Ocimene, XII = Bicyclogermacrene

stituents. Thus, a close correlation of light-induced plastid differentiation and synthesis of volatile terpenes was concluded.

Indications for a light induced accumulation of volatiles are accumulating (Mulder-Krieger et al., 1988). A first report on terpene formation by an illuminated in vitro tissue of *Citrus limonia* did not receive appropriate attention (De Billy and Paupardin, 1971). These cell cultures, however, lost their ability to generate volatile terpenes during subculturing, and only the synthesis of carotenoids was maintained. A special combination of phytoeffectors, as suggested by Agrawal and coworkers (Agrawal et al., 1991), created cultivation conditions leading to the formation and accumulation of volatiles by various *Citrus* sp. It was established that cells of lime, lemon, and grapefruit produced the same aroma chemicals as did the parent plant. *Citrus paradisi* callus generated most of the monoterpenes found in *Coleonéma album* (Fig. 9.5), and further oxidized terpenoic volatiles (Fig. 9.7). As a result, both the physical stimulus light (color and photoperiod) and a suitable selection of phytoeffectors were required. In these experiments light significantly

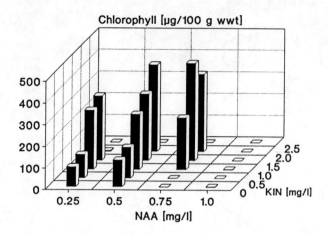

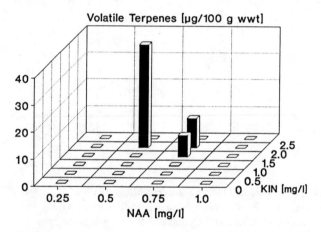

Fig. 9.6. Chlorophyll and sum of mono-and sesquiterpenes of photomixotrophic calli of *Coleonéma album* depended on phytoeffector concentrations (naphthyl acetic acid > 0,75 mg L^{-1} or kinetin > 2 mg L^{-1} were lethal)

affected the quantity, but not the composition of the essential oil (Fig. 9.8). The direct correlation of chlorophyll and terpene contents supported the hypothesis of a chloroplast dependent biosynthesis. Chlorophyll formation, while not being a single essential criterion, was found to be a helpful visual selection marker. A systematic study using *Coleonéma* callus cellus cells showed that a photoperiod of eight hours at 12 000 lux resulted in maximum formation of total volatile terpenes. A further extension to continuous illumination did further increase the concentration of chlorophyll (450 mg kg^{-1}), but not the concentration of volatiles (Reil, 1993). These biosynthetic pathways were maintained stably and reproducibly for more than three years of subcultivation. These results dispel the idea that aroma formation is dependent on morphologically differentiated tissue cultures.

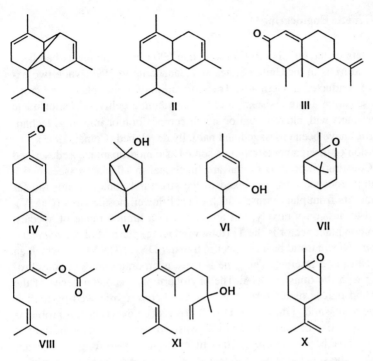

Fig. 9.7. Mono- and sesquiterpenes identified in callus cultures of *Citrus paradisi* (MSL medium, continuous light 3000 lux, 23 °C). I = α-Copaene, II = δ-Cadinene, III = Nootkatone, IV = Perillaldehyde, V = Sabinene hydrate, VI = Piperitol, VII = Pinene epoxide, VIII = Geranyl acetate, IX = Nerolidol, X = Limonene epoxide

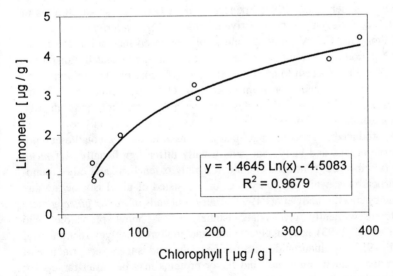

Fig. 9.8. Correlation of chlorophyll and limonene contents of photomixotrophically cultured cells of *Citrus aurantifolia* under varying conditions

9.3.6 Plant Genetic Engineering

According to data of the OECD (1992) more than 600 transgenic plants had been released for field trials in the United States and Canada up to 1993, while two experiments were conducted in Germany. Transformation of edible plants and crops aimed at less sensitive, more resistant, and higher yielding cultivars. Tailor-made agricultural products with altered lipid or starch composition or improved technological properties have been envisaged and partially developed (Comai, 1993).

The expression of foreign genes requires some basic modifications, as described in Sect. 8.2: Constitutive promotors that are functional in all plant tissues and a plant terminator sequence must be linked to the structural gene in a not too far distance. Promotors from plant viruses, such as cauliflower mosaic virus (CaMV), are considered to act constitutively, but are active in a limited range of species. The most common plant vector is the Ti (tumor inducing) plasmid of *Agrobacterium tumefaciens*. This plasmid bears a specific transfer DNA (tDNA) into which an exogenous gene can be inserted. The gene is then incorporated into a plant chromosome along with the transfer DNA. The phytohormone encoding genes of the Ti-plasmid would induce tumor formation and are therefore, replaced by marker genes in chimeric plasmids ('disarmed tDNA'). To reduce experimental problems caused by the large size of the Ti-plasmid, downsized vectors containing the minimum elements have been developed. Other functions necessary for gene transfer, but not located on the small vectors are donated from a separate 'helper' plasmid.

First results on the production of aroma phytochemicals by transformed cells have been reported. Crown-gall tissue of *Anthemis nobilis* (false camomile) and *Agrobacterium* transformed shoot cultures of peppermint species contained essential oils with a composition comparable to that of the oil bearing tissue of the parent plant (Fauconnier et al., 1993; Spencer et al., 1993). The shoot cultures of *Mentha piperita*, for example, were derived using an *Agrobacterium tumefaciens* vector containing the CaMV promoter and stably produced menthol as the major volatile for a period of more than two years. *A. tumefacieus* mediated transformation of strawberry is expected to have potential in producing plants with improved sensory quality upon regeneration from protoplasts (Jain and Pehu, 1992). 'Hairy root' cultures, infected with *A. rhizogenes*, have been suggested for the production of secondary metabolites (Toivonen, 1993). As discussed above, only few impact volatiles are produced in roots of intact plants. Consequently, the volatile components of whole organs and hairy root cells usually differ significantly. *Artemisia absinthium* (wormwood) root cells contained mainly α-fenchene and other monoterpene hydrocarbons, while the hairy root oil consisted of neryl isovalerate and other terpene esters (Kennedy et al., 1993). Hairy root cells of *Panax ginseng* were able to stereospecifically glycosylate isomers of 2-phenyl propanoic acid (Yoshikawa et al., 1993) and to glycosylate and malonylate glycyrrhetinic acid (Asada et al., 1993). A glucuronidoglucuronide of the acid is the sweet principle of liquorice products. Such conjugates and flavor proteins may be easier targets for plant biotechnology. A single-chain gene encoding for monellin, a sweet plant protein, was placed under the control of a fruit-ripening specific, constitutive pro-

motor and, upon transfer to tomato or lettuce, expressed in fruit and leaf, respectively (Penarrubia et al., 1992). The aim of this study was not to produce the swee-
tener biotechnologically, but to enhance flavor and quality of the plant product as
such. The flavor peptide of *Theobroma cacao* (cocoa) has been produced by transformed yeasts, after the cDNA was cloned and sequenced (Mars G.B., 1991).
Transformants of *E. coli* were also investigated and yielded up to 10 mg of peptide
product$\times L^{-1}$.

The current situation is characterized by efforts to find promotors that efficiently drive expression of foreign genes in the plant cell (Ozcan et al., 1993), and
approaches to physically map gene loci related fo flavor formation. Detecting a
single gene among 50 000 plant genes presents quite a challenge. Varietal, seasonal, and processing effects on raspberry flavor have been resolved using multiva-
riate statistics, and the authors believed that diploid berry crops with simple geno-
mes could be screened for the presence of genetic markers using the PCR
chemistry and automated protocols (Paterson et al., 1993). Unknown gene loci
may be identified using restriction fragment length polymorphism (RFLP)-
analysis. The principle is to identify a DNA marker by testing restriction enzyme
digests in hybridization experiments against DNA of nearly isogenic cells with and
without the target gene. This technique has been successfully applied to tag a gene
related to the formation of aroma in rice (Ahn et al., 1992). The obvious problem
of this type of research is the multitude of genes required for the synthesis of one
flavor metabolite, and the considerable number of single volatiles required to
compose a well-balanced plant aroma. There still exists no clear picture of whether
such an enormous amount of foreign genetic material will ever be mastered by a
microbial cell. A part of this discussion continues to revolve around 'the lipoxygenase gene' (Hildebrand, 1992). The positive aroma effects brought about by the
attack of lipoxygenases on unsaturated fatty acids have been described (cf.
Sects. 1.5, 5.3.2, 9.1.1). It was supposed that the fresh top notes that are missing
from processed fruits and vegetables might be restored or even enhanced by the
judicious implementation of a 'lipoxygenase gene'. However, multiple forms of
lipoxygenase and the involvement of further enzymatic activities render this approach more complicated than expected at first sight. The transfer of single plant
genes into microbial flavor producers appears to be a more promising aspect on
the short run. A cinnamyl alcohol dehydrogenase from *Eucalyptus* was examined
by sequence analysis of its cloned gene (McKie et al., 1993). The plant enzyme
was found to be homologous to a range of dehydrogenases. In view of the aroma
relevance of a series of cinnamyl derivatives these findings may gain importance in
enzyme technology (cf. Fig. 5.11). If transformed plant cells will provide the biotechnological basis for an industrial process, remains a matter of speculation. Mitochondrial DNA (mtDNA) of *Sorghum bicolor* was analyzed to detect the extent
of somaclonal variation in cell cultures (Kane et al., 1992). A repeat region was
described that became hypervariable when cells were subjected to protoplast culture. All protoclones were found to differ from each other, from the parental cell
suspension, and from the seedlings in their mitochondrial genome arrangement.
Novel cultivation systems should be designed to exert a selection pressure on the

cells by simulating natural growth conditions, hence reducing the tremendous trend to genetic instability.

9.3.7 Scaling-Up Experiments

Genetic heterogeneity is not the only problem the plant biotechnologist has to cope with. Biological characteristics of plant cells, such as slow growth rates, large cell volume, and intracellular storage sites (cf. Table 4.1) result in specific technical requirements with respect to maintenance of sterility, shear sensitivity, cell aggregation and nutrient and light transfer. Several concepts have been developed to overcome the inadequate stirred-tank batch processes. As rapid growth and formation of aroma compounds were often reported to exclude each other, two stage processes have been considered. Accumulation of active biocatalyst was achieved on full media, and accumulation of product proceeded in minimal media combined with encapsulation techniques. A typical immobilized system uses calcium alginate or polyamide encapsulated cells and maintains cell vitality, but not cell division. A continuous process in a column reactor is the result. Hollow fiber reactors entrap cells between selectively permeable membranes, and nutrient supply and product removal are provided by diffusion along self-forming gradients. Two comprehensive reviews dealt with the recent achievements (Scragg and Arias-Castro, 1992; Doran, 1993; Table 9.14). Multicycle culture techniques (Lipsky, 1992), a bubble column reactor with controlled light intensity (Kurata and Furusaki, 1993), and the revival of the simple rotary drum for the cultivation of suspended (Shibasaki et al., 1992) and differentiated cells (Scragg and Arias-Castro, 1992) illustrate by the diversity of technical designs that the present stage of development is characterized by a strong need for more basic research. As the most complex biocatalyst, plant cells should benefit from technological advances made in other fields of biotechnology. However, the in vitro cultivation of plant cells on the basis of data collected from bacterial bioprocesses should be discontinued. Considering the particular physiology of plant cells in bioengineering by developing improved gas, light, or substrate supplies and in situ product separation will assist in establishing higher yielding laboratory-scale plant cell cultures.

Table 9.14. Recent developments in the larger-scale cultivation of in vitro plant cells

---

Self-immobilized plant cells
Fluidized bed reactor
Bubble-column reactor
Helical and anchor impellers
Embryo culture and spin filter reactor
Phototrophic culture and optical fiber reactor
Hairy root culture and droplet phase reactor

---

10 Bioprocess Technology

Type, stability, morphology, and surface properties of biocatalysts used for aroma generation (except plant cells) do not fundamentally differ from those of biocatalysts used in established processes of food and fine chemicals biotechnology. Similarly, the composition, rheology, coalescence and foaming properties of the nutrient media and the physical process parameters of present aroma yielding processes are comparable to data gathered from a variety of prokaryotic and eukaryotic cultivations. As a result, typical design parameters have to be considered for the layout of the reactor, scale-up, and optimization:

- Mass transfer in submerged cultivations has been well studied and proceeds in a three-phase system (microorganisms, liquid bulk phase, gas bubbles). Solid state and immobilized enzyme reactors have been much less investigated.
- Heat transfer, depending on the metabolic waste heat and the biological optimum temperature, can have a negative or positive sign. Hot spots in immobilized or self-aggregated systems are difficult to control.
- Kinetics of the biochemistry (growth, substrate and product concentrations) should be devoid of inhibitory or inactivation effects.
- Physico-chemical parameters (viscosity, equilibria, interfacial interactions) can be crucial and pose optimization problems particularly in integrated processes.

Some aspects of improved medium composition and fluid dynamics have been reviewed recently (Schügerl et al., 1990).

10.1 Bioreactor Design

Most industrial-scale bioprocesses are batch cultivations. Nonlinear changes of cell number, viscosity, mass transfer capacity, and product concentration are typical. Time dependent steady states impede constant operation and technical optimization. Mycelial growth of fungal organisms, the most attractive group of microbial aroma producers (cf. Sect. 5.4), affects the apparent viscosity in relation to mycelium density, pellet formation, fluid velocity and proportion of emulsified air. Highly viscous cultures require high energy input and suffer from low oxygen transfer rates. The classical, impeller agitated stirred tank reactor does not meet all requirements of more demanding cell types. High transfer rates and homogeneous distribution of cells and nutrients remain illusive, particularly on the larger production scale. Good mixing properties in the surroundings of the impeller blades are usually not paralleled by sufficient blending of the bulk fluid. The circulating effect of the ascending gas bubbles must be mechanically supported at higher cell densities.

The major disadvantage of stirred tanks is the impossibility to scale-up geometrically similar reactors while simultaneously maintaining heat and mass transfer parameters. In a tower loop reactor, mixing, pumping, and aeration are functionally separated, resulting in a constant and identical passage of all fluid elements through the different zones of the system. The theoretically formed plug profile is disturbed by back mixing effects in turbulent regions. However, fairly high productivities and good specific productivities (related to cell mass) can be achieved with moderate specific power input (Schügerl, 1993). Local measurement can help detecting nonuniform distribution of nutrients in the reactor. The performance of bioreactors for aroma production by fungi or recombinant organisms is still difficult to evaluate.

10.1.1 Mode of Operation

Some operational problems of simple bioreactors can be overcome by more refined modes of operation. A fed-batch cultivation of brewers' yeast resulted in increased biomass and fermentation activity (Strel et al., 1993). Yeasts were cultivated with linear (Mignone and Rossa, 1993) and parabolic gradient feed of nutrients (He et al., 1993). The parabolic feed method was recommended for species that show the *Pasteur* effect (reduced glucose utilization by fermenting yeast at elevated pO_2; the *Crabtree* effect refers to a repression of respiration by high carbohydrate concentrations), because the supplementation of the carbon source was coinciding with the growth kinetics.

Two stage operation was advantageous in a process where enzyme induction was followed by transient product formation (Cubarsi and Villaverde, 1993). A stable product concentration in the outgoing stream allowed a continuous downstreaming. If no optimal residence times can be found, better performance may be achieved in a continuous or even batch cultivation. Sequential reactions are desirable in applications where the pH optima or other parameters of two associated steps differ significantly (Byers et al., 1993). Artificial compartments can be created using membranes or reactor cascades. Two stage cultivations were also useful to reveal interdependencies of mixed populations. Synergistic effects of *Acetobacter xylinum* and *Schizosaccharomyces pombe* ('tea fungus') are involved in the traditional conversion of sweetened black tea to 'Kombuchta', a beverage with presumed probiotic activities. Sensory deficiencies are common. Separation of the pure cultures and growth of the yeast on the cell free medium of the first *Acetobacter* fermentation resulted in a palatable, cider-like product and allowed to track down the origins of the volatile flavor compounds (Table 10.1). Typical yeast volatiles increased during the fermentation (Fig. 10.1). Maillard products, formed during sterilization of the tea medium and the cell free *Acetobacter* medium of the first stage, were present in high amounts at the beginning of the second

Table 10.1. Volatiles of a low alcohol beverage upon two-stage fermentation of black tea

Compound [a]	Origin [b]
2-Methyl-3-butan-2-ol	S
2-Methylpropanol	T,S
2-Pentanol	T,S
2-Methyl-2-butenal	S
3-Methylbutanol	S
3-Methyl-3-butan-1-ol	S
Acetoin	S
3-Methyl-2-butan-1-ol	S
3-Hydroxy-pentan-2-one	S
Ethyl lactate	S
Linalooloxides (four isomers)	T
Acetic acic	A,S
Benzaldehyde	T,S
2,3-butandiol	S
Linalool	T
5-Methyl-dihydrofuran-2(3H)-one	S
α-Terpineol	T
3-Methylthiopropanol	S
Benzylalcohol	T,S
2-Phenyl ethanol	T,S

[a] in order of elution on CW20 M
[b] A: *Acetobacter*, S: *Schizosaccharomyces*, T: black tea

fermentation stage (Fig. 10.2). Yeast activity decreased the amounts of the furanoid compounds. A maximum concentration of furfurol during the first day of
cultivation coincided with lowest furfural and hydroxymethylfurfural concentrations, indicating overlapping redox reactions of the yeast.

10.1.2 Fluidized Bed Reactor

Advantages claimed for the use of either immobilized cells (cf. Sects. 2.2, 2.3.5,
2.10) and enzymes (cf. Sect. 7.1.5) have been discussed above. Few of the claims
have been fully substantiated over the years, and some may not even be entirely
justified. Enhanced mechanical and metabolic stability of immobilized cells,
however, have been widely observed. This gains particular importance in fluidized
bed reactors, where an external gas stream is introduced into layers of encapsulated cells (Bejar et al., 1992). Yeasts are often used as model organisms which

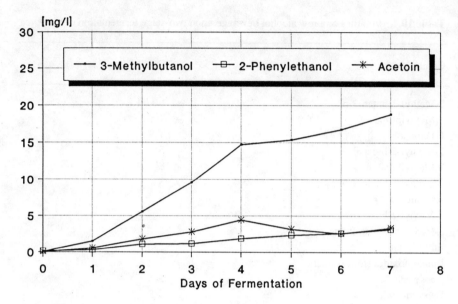

Fig. 10.1. Volatiles of *Schizosaccharomyces pombe* during fermentation of an *Acetobacter xylinum* pregrown black tea medium

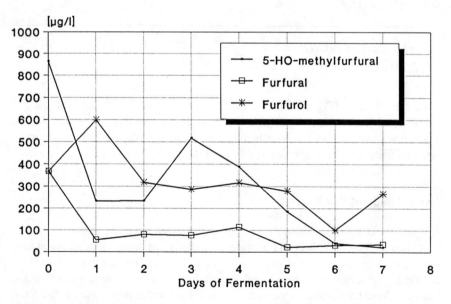

Fig. 10.2. MAILLARD volatiles decreased during growth of *Schizosaccharomyces pombe* in a heat sterilized medium

facilitates a transfer of results to aroma yielding processes. A fluidized bed process using *Zymomonas mobilis* produced ethanol as the main volatile (Weusterbotz et al., 1993). High cell concentrations resulted in short residence times and in washout conditions for contaminants. A semi-technical two stage cascade was operated stably under nonsterile conditions. Problems of particle flotation and gas logging were avoided by a concurrent fluidization from top and bottom, and by fluid removal through a side port. (Gilson and Thomas, 1993). Ongoing work on ethanol formation indicated that the continuous aeration of a fluidized bed could be favorably combined with an integrated recovery of volatile products, hence alleviating product inhibition (Nguyen and Shieh, 1992).

10.1.3 Application of Membranes

Membranes as an integrated part of a bioreactor are often preferred over conti-centrifugation for recycling cells from the effluent of the vessel back to the biore-action (Tin and Mawson, 1993). Methods to optimize ethanol production in such a system have been transferred to other volatile target compounds (Warren et al., 1992). Propanoic acid with its strong sensory and mycostatic properties is among the first examples of aroma applications (Colomban et al., 1993). Product concentrations of 40 g×L^{-1} were reported for a recycle cultivation of *Propionibacterium acidipropionici*. The excess biomass was utilized as starter culture in cheese manufacturing. The compartmentalization of the bioreaction by membranes can also help improve yields by allowing diffusion of inhibitory compounds and transport to a down-streaming unit (Markl et al., 1993). In more basic studies, diffusional limitations and solute concentrations were examined using gel immobilized yeast in a membrane reactor (Debacker et all., 1992). The reactor, a gel layer confined by membranes, was flanked by two mixing chambers; through one chamber substrate was pumped, the second measuring chamber indicated the chemical situation at the surface of the gel layer. By extrapolating to the center of the gel bead and modelling of the data, a tool to study the physiology of the immobilized cells was provided. The aerobic degradation of xylene was studied in a reactor, in which a cell layer was grown on a gas permeable membrane (Debus and Wanner, 1992). Oxygen was supplied through the silicone membrane. This principle could also be used to continuously supply aroma generating cells with oxygen and volatile metabolic precursors. Hollow fiber reactors can afford high cell densities. Immobilized cells of *Lactococcus lactis* were cultivated on glucose or lactose, and the formation of volatile products was monitored (Smith et al., 1993). The pH of the medium strongly affected the product profile in the presence of citrate. The overall products resembled those of a carbohydrate limited culture at low dilution rates. Spirally wound fibers or flat sheets were used to immobilize high volumetric activities of either lipase (Garcia et al., 1992) or *Propionibacterium acidipropionici* (Lewis and Yang, 1992a). The lipase was immobilized by adsorption on microporous polypropylene and used to hydrolyse butterfat as a flavor ingredient. The production of propanoic acid in a scalable system opens alternative ways to an industrial application.

10.1.4 Solid State Reactor

The renewed interest in solid state reactions has been stimulated by universal applicability, simplicity of design, and ease of operation. Solid state processes have been applied preferably to the production of enzymes (Table 10.2). Yields often exceeded those of submerged cultures or of fluidized bed cultures (Kuhad and Singh, 1993; Matsuno et al., 1993). Flavor applications can be easily envisaged (cf. chapter 7). Static processes have also performed superior for the chiral transformation of ketones (Matsuno et al., 1993), and even for the de novo synthesis of secondary metabolites (Pandey, 1992). In view of the particular suitability for the cultivation of the pseudotissue-forming higher fungi (Ortega et al., 1993), solid state systems should no longer be neglected in aroma applications (cf. Sects. 5.4, 8.2.4). Less organized aggregates or well developed fruiting bodies can be grown in these systems. A series of key cultivation parameters were identified (Table 10.3). Steep gradients indicated that heat and mass transfer were often far

Table 10.2. Enzymes from solid state bioprocesses

Enzyme type	Biocatalyst	Reference
Pectinase	*I.*	Anthier et al., 1993
Amyloglucosidase	*Aspergillus niger*	Gowthaman et al., 1993a
Cellulases	*Penicillium citrinum*	Kuhad and Singh, 1993
'Lipolytic activity'	*Penicillium candidum*	Ortiz-Vasquez et al., 1993
Acid protease	*Aspergillus niger*	Villegas et al., 1993

Table 10.3. Key parameters of solid state bioprocesses

Parameter	Reference
Pressure drop across the bed	Auria et al., 1993
Type of solid support	Christen et al., 1993
O_2 and CO_2 gradients	Ghildyal et al., 1992
	Gowthaman et al., 1993b
Particle size and substrate load	Gumbirasaid et al., 1993
Headspace pCO_2	Kerem and Hadar, 1993
N limitation	Larroche and Gros, 1992
H_2O activity and content	Larroche et al., 1992
Heat generation, inhomogeneity	Lonsane et al., 1992
Humidity and temperature of air	Sargantanis et al., 1993

from optimum, and steric limitations by the volume of the support materials were inevitable. However, if a critical thickness of the bed is not exceeded, high cell densities cause aggregated, tissue-like growth that, in turn, can promote product formation. The present front of research is represented by studies following the catabolism of radiolabeled glucose and lignin (Kerem and Hadar, 1993). Compared to submerged systems, less information on scaling-up is available; consequently, commercial processes are largely based on empirical approaches.

10.1.5 Alternative Reactors

The continued study of engineering parameters of airlift reactors (Russell et al., 1994) and their application for the bioconversion of terpenoids (Roy et al., 1992) have confirmed their usefulness in cultivating shear sensitive cell types. Gas stripping of volatile solutes and the coupled aeration and fluid circulation present limitations. Gaseous substrates were metabolized in a pressurized reactor (Pohland et al., 1992; cf. Sect. 2.2). Immobilized yeasts (Abraham and Surender, 1993) and films of *Aspergillus niger* (Sakurai and Iwai, 1992) were grown in systems with rotating elements. The performance of aroma producing yeasts or fungi in such bioreactors has not been studied, not to mention the inavailability of comparative data.

10.2 Process Monitoring

An optimal bioprocess must rely on accurate and real time information on relevant parameters. Some characteristic values can be measured directly (*on line*), other measurements need low dead volume and sterilizable filter modules to analyze cell free medium (cf. Sect. 4.3). Data conditioning, analysis, modeling and the implementation into feed-back control and regulation has been based on personal computers and combined with expert systems. Fuzzy logic is applied to handle imprecise data and to use the operator's experiences (Aarts et al., 1990).

A fundamental parameter is the time dependent change of the biomass. Higher cell densities and complex cell morphologies posed considerable problems in attempts to measure cell growth *on line*. A solid state culture of *Rhizopus oligosporus* was monitored by measuring changes of electrical capacitance (Penaloza et al., 1992). Similarly, gas development during fermentation of bread dough was measured on a dielectric basis (Ito et al., 1992). Metabolic changes that occurred in cultures of *Corynebacterium glutamicum* were detected *on line* by concurrently measuring and comparing dissolved oxygen, the main contributor to the redox potential, and overall redox potential (Kwong et al., 1992). On this data basis, the end of the lag phase and substrate exhaustion were identified. Impeller rheometer (Olsvik and Kristiansen, 1992) and image analysis (Thomas, 1993) improved the characterization of cultures of filamentous organisms.

Spectroscopic methods, such as near infrared spectroscopy, were applied to measure *on line* the solute and biomass concentrations in lactic fermentation (Vaccari et al., 1993). Volatile flavors can be measured *on line* by membrane inlet mass spectrometry (Heinzle, 1992; Lloyd et al., 1992; Lauritsen and Lloyd, 1994). A polymer membrane was the only separation between the bioproduct phase and the vacuum system. While in recent years there have been impressive sensor innovations, little work has been devoted to specific volatile chemicals, with the exception of ethanol. Semiconductor based ethanol sensors have been developed for *on line* measurement of the course of concentration in sake brewing (Oishi et al., 1992; Samuta et al., 1992). Aseptic sampling in such procedures is usually carried out by membranes, or by inorganic filter materials (Picque and Corrieu, 1992). Crossflow operation can be used to increase throughput or to decrease the size of the active filter area (Ohara et al., 1993). Slow equilibrium adjustment, dilution requirements, and time of sample transfer render most of these systems 'quasi on line' only. Separation by microcentrifugation does not make a difference in this respect (Turner et al., 1993). Therefore, product oriented measurements of aroma chemicals via the gas phase using rapid gas chromatography or total carbon analysis are not only more relevant, but reflect real time behavior better than analysis of less volatile indicator compounds.

Specific optical and electrical principles were used to develop measuring probes for certain applications. Increased selectivity can be obtained at the expense of heat lability by implementing biomolecules, such as enzymes, antibodies, or DNA-sequences into the measuring device. The transducer most frequently used was an O_2-electrode, and the analytes were carbohydrates, lactate, ethanol, and various amino and organic acids.

10.3 Downstreaming Volatile Flavors

Downstream processing is a term covering isolation, concentration, and purification of a bioproduct. The present industrial standard of aroma biotechnology, batch fermentation with *off line* distillation (or extraction), requires the handling of large volumes of aqueous solutions. The well-known double log correlation of product concentration and selling price indicates that the usually low concentrations of the aroma compound in the aqueous bulk phase make conventional downstreaming a major cost factor. Volatile flavors, like pharmaceuticals, need to be devoid of any bioactive contaminants, otherwise the sensory quality would be deteriorated. For example, bioprocesses at elevated temperatures or sterilization conditions result in the formation of MAILLARD products with undesirable odor notes (Fig. 10.2). At least four further reasons suggest an in situ removal of bioflavors using advanced separation techniques:

– Loss of product via the waste air stream due to volatility,
– biochemical instability of product in the presence of the producing cell,

- inhibition phenomena, and
- nonstationary product concentration in conventional batch processes.

The necessity of developing more specific in situ techniques applicable under sterile conditions and in the immediate surroundings of the biocatalyst has now been recognized (Freeman et al., 1993). Volatile flavors are among the bioproducts that could gain most from these novel approaches (Abraham et al., 1993).

10.3.1 Degradation of Volatiles

Many bacteria and fungi accept a broad range of carbon substrates, including flavor compounds (cf. Sect. 5.5) and even xenobiotics. The degradation of chlorophenols by *Pseudomonas* (Kiyohara et al., 1992), by coryneform bacteria (Kramer and Kory, 1992), and by anaerobic cultures (Mohn and Kennedy, 1992), and fungal ring fission and oxidation reactions have been biochemically investigated from ecological views and can illustrate the metabolic flexibility (Wright, 1993). Microbial oxidoreductases for the generation of volatile flavors were mentioned above (cf. chapter 6, Sect. 7.4). One of the specific problems of aroma biotechnology is to accumulate products with an intermediate stage of oxidation. Advanced functionalization of primary products will lead to reduced volatility and flavor value (cf. Sect. 1.1). If biochemical measures of aroma accumulation fail or prove insufficient, technical measures need to be taken.

10.3.2 Inhibition Phenomena

Inhibition of enzymes by high substrate concentrations, low concentrations of structural analogues, and by the product provides living cells with sufficient sensitivity to adapt to changing environments, and provides the biotechnologist with reasons for continuous processes and integrated downstreaming units. If active cellular transport is lacking, the product may be unable to diffuse rapidly out of the microenvironment producing substantial inhibition, though its concentration, averaged over the whole system, remains negligible. The inhibitory effects of ethanol and free fatty acids on yeast and other organisms have already been discussed (cf. Sects. 2.1, 5.3.1, 5.3.5). Yeasts pregrown on higher levels of ethanol acquired increased ethanol tolerance (Lloyd et al., 1993). Toxic functions of ethanol and fatty acids on *Saccharomyces cerevisiae* were attributed to passive proton influx (Stevens and Hofmeyr, 1993). The fermentation rate was inhibited in a way indicating a correlated effect. Cells in the first phase of cultivation often exert higher sensitivity toward fermentation acids and other products (Yamamoto et al., 1993). For a detailed discussion of alternative biochemical explanations the reader is referred to Russell (1992).

The cytotoxicity of plant monoterpenes reflects inhibitory activities (cf. Sects. 2.3.1, 5.2, 6.1, 9.3.4). A recent study compared water solubility and phytotoxicity of monoterpene hydrocarbons and oxigenated analogues (Weidenhamer et al., 1993). Many monoterpenes were phytotoxic in concentrations < 100 mg×L^{-1}, well below the saturated aqueous concentration. These results agree with data obtained for numerous microbial and plant cell cultures (Berger, 1994). Several phenols, such as 4-hydroxybenzaldehyde and vanillin stimulated fungal growth at low concentration, and decreased growth at levels > 5 mM (Cai et al., 1993). As for *Pseudomonas fluorescens* (Bare et al., 1992), the intermediate formation of the phenylcarboxylic acid may characterize the degradative pathway in fungi. An obvious conclusion is that formation of vanillin or, in comparable cases, of other volatile flavors will only be achieved by catabolically blocked mutants. For example, mutants of *Pseudomonas aeruginosa* were isolated that were incapable of removing the 3-methyl function of citronellic acid (Hector et al., 1993). These mutants were also deficient in the utilization of structurally related monoterpenes and could, therefore, be useful in biotransformations. The situation where a saturated aqueous concentration of product during a bioprocess results in separation by crystallization as in the case of L-tryptophan will remain a rare exception (Azuma et al., 1993).

10.3.3 Solid/Liquid Separation

Most volatile metabolites are synthesized to be lost and are accordingly excreted into the nutrient medium. To obtain a cell free medium for facilitated further downstreaming, gravitational sedimentation of *Saccharomyces cerevisiae* cells was applied (Maia and Nelson, 1993). A larger scale and continuous production of ethanol was envisaged using a calcium alginate filter in the overflow for improved cell recycle. Continuous throughput centrifuges may present a technical alternative for large scale processes, but suffer from high investment and operation costs. Shear sensitive cell types can be completely separated at about 50 g; prokaryotics withstand much greater forces.

More popular are the various membrane based filter systems. One step tangential-flow filtrations avoid the formation of a cell deposit on the active membrane by shear/lifting forces and can be applied in submerged cultivations of bacteria (Ishizahi et al., 1993). Sterilizable filtration membranes have stimulated the development of integrated micro-filtration systems (Groot et al., 1992). Continuous fermentation of ethanol by yeast can again serve as a model process for the generation of a volatile compound: The performance of a 2-µm pore size, stainless steel filter was effected by agitation speed and density of the yeast cells (Chang et al., 1993). Best operable cell concentrations of 90 to 150 g×L^{-1} were similar to those of an external membrane cell recycle. Biomass of the pellet forming fungus *Tyromyces sambuceus* was recycled using a 100 µm pore size polyester membrane (Kapfer et al., 1991). The fed-batch culture was maintained for 130 days. Intermittent feeding of castor oil resulted in elevated concentrations of 4-decanolide (cf. Sect. 5.4.3, Fig. 10.3). Membranes are even penetrating the field of organic synthesis: A Ru(III) EDTA complex, useful in hydroformylation and carbonylation reactions, was recycled using a cellulose acetate ultrafiltration membrane

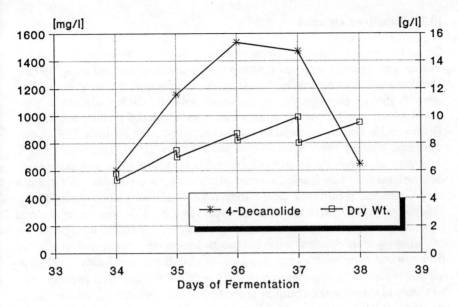

Fig. 10.3. Fed-batch culture of *Tyromyces sambuceus* with biomass recycle (Kapfer et al., 1991)

(Halligudi et al., 1992). An almost complete recovery of the small catalyst molecule was reported.

A second solid/liquid separation is required, if an intracellularly stored target compound has to be liberated from the biomass. As the separation quality of membranes or centrifuges strongly depends on particle size distribution, the choice of the method of cell disintegration critical. High pressure homogenizers with advanced valve designs and high speed mills are upscalable alternatives. The combination with lytic enzymes can result in larger cell wall fragments and facilitates the separation step. The enzyme costs are, however, a limiting factor.

10.3.4 Electrodialysis

Passing the border to nonvolatile flavors with increasing molecular weight, carboxylic acids represent aroma chemicals amenable to electrochemical separation. In a typical experiment, acetic and propanoic acid were recovered in situ from cultures of *Propionibacterium shermanii* by electrodialysis (Zhang et al., 1993). The growth inhibitory effect of propanoic acid could be alleviated (cf. Sect. 10.3.2). Fouling of the membranes and interference of inorganic ions were among the problems observed (Boyaval et al., 1993; Weier et al., 1992). It is still unclear whether the salt levels of common medium components, such as sweet whey permeate, will permit economic larger-scale processes.

10.3.5 Reactive Extraction

Carboxylic acids can also be separated directly from the nutrient medium by adding an appropriate counter ion such as a tertiary amine (Lewis and Yang, 1992b). The propanoic acid present in the extractant was stripped off using an alkaline solution, thereby also regenerating the amine solution. Further advantages mentioned were better pH control of the process and shortened product purification. The reversible binding of acetaldehyde was achieved using nontoxic TRIS buffer (Duff and Murray, 1992; Shacharnishri and Freeman, 1993). The biocatalysts in these processes were *Pichia pastoris* and alcohol dehydrogenase and alcohol oxidase isolated from *Candida* sp., respectively. The *Pichia* cells were cultivated in a semicontinuous, pressurized closed loop reactor (cf. Sect. 10.1). Acetaldehyde yields of up to 130 g$\times$L^{-1} in four hours were reported. This was one of the best productivities of an aroma generating bioprocess ever achieved. The regeneration of carbonyls from chemically more stable hydrazone and oxime derivatives was demonstrated using baker's yeast in organic solvent (Kamal and Reddy, 1992). The various processes have in common the deliberate use of a chemical agent in order to transiently modify the chemical properties of the final product. According to current legislation an aroma compound manufactured in such a way cannot be classified as 'natural' (cf. Sect. 3.2). Disregarding the legal aspect, reactive extraction due to its efficiency and compatibility with a continuous process is one of the most intriguing options for the downstreaming of aroma compounds.

10.3.6 Two Phase Extraction

The lipophilic nature of most aroma chemicals has induced the idea of co-cultivating a second, water immiscible phase as a synthetic accumulation site (sf. Sect. 9.3.4). (In more correct terms, the 'second' (liquid) phase is the fourth phase in standard submerged cultures.) Among the adverse effects of the second, organic phase were reduced cell viability, and coalescence and foaming problems. Technical and biochemical measures to partly overcome the current drawbacks have been reviewed (Vansonsbeek et al., 1993). Experience gathered from the extractive production of ethanol, *n*-butanal, and *i*-propanol could be transferred to aroma biotechnology (Araki et al., 1993; Davison and Thompson, 1993; Jones et al., 1993). In Europe, the use of extraction solvents has been regulated by the Commission (Guideline 88/334 of 13.6.1988). Many authors have preferred the use of immobilized cells and longer chain fatty alcohols as extraction solvent. Oleyl alcohol did not interfere with growth and vitality of the fungus *Poria cocos* in standard batch cultures. The 4-octanolide producing fungus *Polyporus durus* was cultivated in the presence of an excess of precursor triacylglyceride (Fig. 10.4). During metabolism of a proportion of the second liquid phase, the same substrate dissolved its catabolite lactone. Improved dispersion of the extractive oil was obtained by adding an emulsifier to the aqueous phase, resulting in enhanced lactone formation up to peak concentrations of 500 mg$\times$L^{-1}. The oil phase showed an experimentally

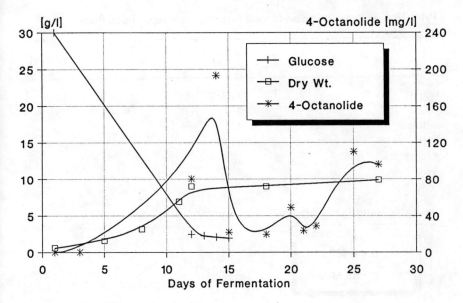

Fig. 10.4. Extractive fermentation of *Polyporus durus*, the precursor triacylglyceride serving as an extraction solvent

determined distribution coefficient near six for 4-octanolide. Six to ten carbon fatty acids were converted to the corresponding methyl ketones by *Penicillium roquefortii* (Creuly et al., 1992; cf. Sect. 5.3.3). Technical isoparaffin (mainly tetradecanes) was used to isolate about 150 g of methyl ketones$\times$L^{-1} spore suspension.

Aqueous two phase systems, stabilized by industrial dextran or other modified starchy polymers and polyethylene glycol, have increased the yields of acetone/butanol bioprocesses (Kim and Weigand, 1992). Cell and product partitioning can be strongly affected by biosurfactants produced during the bioprocess (Drouin and Cooper, 1992). Continuous process operation, immobilized or recycled biocatalysts, and reduced product inhibition are the recurring key words (Jarzebski et al., 1992). In view of the inferior processing parameters it might rather be the increasing impact of ecological and safety considerations that will smooth the way into aroma applications.

Supercritical fluids show properties intermediate between those of gasses and liquids that can be controlled by either temperature or pressure, or by adding entrainers. Carbon dioxide, due to convenient critical parameters, nontoxicity, chemical inertness and cheapness, has found a significant number of food applications including the extraction of volatile flavors from hop and essential oil plants (Table 10.4). More recently, 5-alkanolides were separated from anhydrous milk fat (Rizvi et al., 1993). If an aroma compound was generated in a submerged culture, a preconcentration on an adsorbent was required to discontinuously recover the trapped volatile(s) (Kapfer et al., 1990). Model experiments using 4-decanolide, one of the first commercial bioflavors (cf. Sect. 5.4.3), have been conducted to

Table 10.4. Critical parameters of some frequently used supercritical fluids

	T [°C]	p [Mpa]
CO_2	31.3	7.39
H_2O	374	23.0
n-Butane	152	3.8
NH_3	132.5	11.4
Xe	16.6	5.9

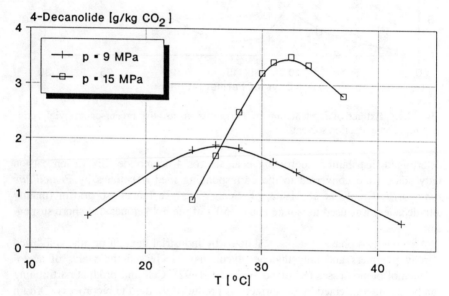

Fig. 10.5. Isobaric solubility of 4-decanolide in carbon dioxide (Kapfer et al., 1992)

measure statically the solubility at different temperatures and pressures (Fig. 10.5). Adsorbed 4-decanolide was almost completely recovered from XAD-2 as the polymer carrier (Fig. 10.6). A high throughput of carbon dioxide was required to achieve this result, but in fairness to supercritical fluids, extraction using a conventional liquid is often likewise incomplete. In parallel sets of extraction experiments, only 60% of the adsorbed 4-decanolide was desorbed from chromosorb, and less than 2% was desorbed from activated charcoal.

Another novel approach, liquid membrane extraction, attempts to combine liquid/liquid extraction with membrane separation (Eyal and Bressler, 1993). Large specific interfacial areas were created in emulsified liquid membranes (Scholler et al., 1993). A typical membrane phase for the separation of ionic species, such as lactate, consisted of Span 80 dissolved in n-heptane/paraffin, a tertiary amine car-

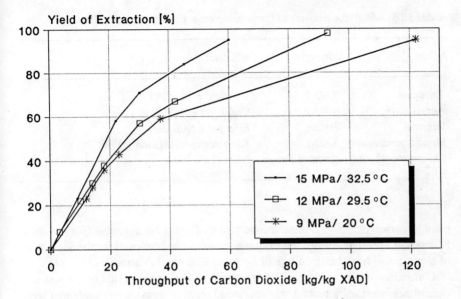

Fig. 10.6. Extraction of 4-decanolide loaded XAD-2-resin by carbon dioxide (Kapfer et al., 1992)

rier, and aqueous carbonate in the internal phase. Low viscosity polysiloxanes broadened the range of separable compounds. These polymer carriers could be chemically functionalized with specific nonchiral or chiral host molecules to achieve high selectivity and reversible binding. An elegant variant are charged microemulsions that are added to a product containing bioreactor. One technique is based on the subsequent addition of oppositely charged gas aphrons (microfoams) to float the colloidal liquid to the medium surface. The large oil/water interface of colloidal liquids (about 3000 m^2 per liter were calculated) is transformed into rapid rates of solute extraction. Thus, direct mechanical separation, for example by using ceramic filter, represents an alternative technique. Nonpolar compounds such as chlorobenzenes were successfully separated in first laboratory scale experiments.

10.3.7 Solid Phase Extraction

The step from absorption of volatiles to adsorption is made using inorganic or organic polymeric solids. Adsorption as a physico-chemical process marks not only the beginnings of chromatography, but is nowadays accepted standard to accumulate analytes in gaseous samples of environmental and aroma chemistry and to enrich pesticides for the analysis of drinking water (Font et al., 1993). In technical and biotechnical application considerable disagreement appears to exist with respect to the choice of the most appropriate adsorbents (Table 10.5), and few com-

Table 10.5 Adsorptive recovery of flavor compounds from the aqueous phase

Solute	adsorbent	reference
Thiopenes	XAD-7	Buitelaar et al., 1993
Amino acids	polystyrenes	Casillas et al., 1992
Geosmin	Zeolite Y	Ellis and Korth, 1993
Monoterpenes, etc.	XAD-2	Klingenberg and Hanssen, 1988
Chlorophenols	activated carbon	Najm et al., 1993

parable studies of the adsorptive recovery of bioflavors are available (for example, Kühne and Sprecher, 1989). Ion exchange resins were employed for the separation of L-lactic acid from batch culture of *Lactobacillus casei* (Vaccari et al., 1993).

Continuous *on line* solid phase extraction with concurrent reflux of the aqueous eluate back into the bioreactor was applied to recover aroma compounds that were not volatile enough for gas stripping (Krings and Berger, 1994). A preceding *on line* filtration is necessary to avoid blocking of the adsorbent column by biomass particles. Switching between two or more adsorbent columns gives time to desorb the product from the loaded column either thermally or by using a recyclable organic solvent. The possibility of using solvent gradients moves this technique near to liquid chromatography. A more detailed comparison of adsorbents showed that activated carbons extracted maximum amounts of volatiles from an aqueous phase, but solvent desorption was < 50% (Krings et al., 1993). Styrene/divinylbenzene based macroporous resins and zeolite likewise adsorbed high amounts of lipophilic compounds, but could also be almost completely desorbed. If aroma compounds and further lipophilic compounds with similar adsorption characteristics are produced in the course of the process, co-adsorption occurs and decreases the time of breakthrough of the target compound. Experiments comparing simulated yeast extract medium and genuine fermentation medium with deionized water showed that the effects of the matrix on the performance of a low pressure downflow fixed bed were much less pronounced than expected (Fig. 10.7). A complete recovery depends on a sufficient amount of adsorptive material and operation at flow rates that permit equilibrium formation. Both parameters can be derived from lab-scale experiments and calculations based on the MTZ (mass transfer zone) or the LUB (length of unused bed) model. In contrast to a common prejudice, no measurable decrease of adsorptive capacity was found after more than 20 adsorption/desorption cycles. Careful conditioning of the resins and moderate desorption rates contributed to this finding (Krings and Berger, 1994). The most polar aroma compounds (furanones, phenols, etc.) compete for the adsorptive sites with excess water resulting in generally poor recovery. Furthermore, solid phase extraction is not applicable to continuous biotransformations with stationary concentrations of lipophilic precursor.

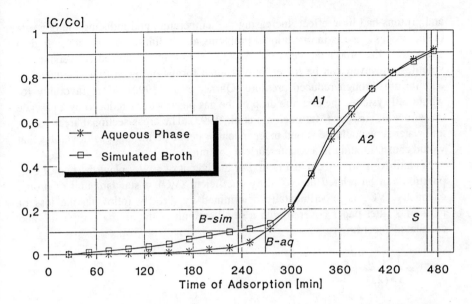

Fig. 10.7. Adsorption of 4-decanolide [189 mg/l] onto 5 g Lewatit 1064: 2.6 liter of aqueous phase or 2.63 liter of simulated fermentation broth per hour, respectively; B = breakthrough level 10%, S = saturation level 90%, symmetry factor = areas A1/A1+A2

10.3.8 Gas Extraction

The operation of conventional stirred tank, loop, or airlift reactors inevitably leads to large amounts of volatiles (supplied to or generated in the reactor) being lost to the waste air stream. A closed system, operated with a recycling O_2 sparge and a CO_2 scrubber, solved the problem for the biodegradation of 1,2-dichloro ethane in waste-waters (Dos Santos and Livingston, 1993). Conversely, gas stripping has been favored by various measures and utilized to separate ethanol from solid phase fermentations. Strictly anaerobic fermentations of *Clostridium acetobutylicum* were stripped with oxygen free nitrogen. Immobilized yeast was cultivated in two column reactors that consisted of two gas/liquid flow sections: a first – cocurrent enricher-accumulated ethanol in the CO_2 phase, and a second – countercurrent stripper-removed volatiles together with the carrier CO_2 from the spent medium (Dale et al., 1985). Regenerated CO_2 was recirculated. Stripping gas flow rate and dilution rate of the (immobilized) cells were identified as key parameters improving productivity (Mollah and Stuckey, 1993). With respect to continuous product formation and finite partition coefficients between gas and liquid phase, it cannot be expected to reach zero levels of products in the reactor. However, concentrations of volatiles below inhibitory levels are a realistic developmental target. Chemical parameters were studied in a model liquid culture medium (Lethanh et al., 1993). Volatility of aroma compounds was increased by inorganic salts and decreased in the presence of fatty acids, while glucose, amino acids, microelements and protons had little effect. Increasing the temperature and reducing the pressure of the process results in operational problems and is limited by the physiological changes of the producing cells (cf. Sect. 2.2). Enzyme catalyzed conversions of volatiles, however, can be forced into product formation by shifting the chemical equilibrium through reduced pressure (Ohrner et al., 1992). More favorably for whole cell systems, the pressure drop of the gas phase can be reduced by a cascade crossflow air stripper (Mertooetomo et al., 1993). The cross-sectional areas for air and water flow could be varied independently, hence permitting gas flow rates that would cause flooding in a countercurrent column.

'Volatility' is not a technical term. For the layout of a pilot plant the rate of evaporation can be related the volatility coefficient (VC), a standard unit of aroma chemistry. VC_m is experimentally determined by directly following the loss of mass of a filter paper covered with a small amount of the aroma compound, and calculated according to

$$VC_m = \frac{1*100}{t*(2a)}$$

with l = loss of mass [%], t = time [h] and a = area of capillary extension [cm^2]. By multiplication with the mass of the compound, a generalized mass independent measure (VC) is obtained. Some values of well-known flavors show the large bandwidth of the coefficients (Table 10.6).

After the nutrient medium as been stripped by a carrier gas stream, another phase transition is required to obtain the pure volatiles. This can be achieved by

- membranes (cf. Sect. 10.3.9),
- adsorption (cf. Sect. 10.3.7),
- condensation, or
- absorption.

Table 10.6. Volatility coefficients (VC) of some aroma compounds

Compound	VC
Vanillin	0.0037
4-Decanolide	0.11
2-Phenyl ethanol	1.13
Linalool	8.48
Benzaldehyde	27.56
3-Methylbutanol	80.65
n-Butanol	302.3
Ethanol	612.9
Ethyl acetate	3177

Adsorption (and accumulation) of volatiles on the solid interface is characterized by high diffusion rates and can be easily adapted to any batch process. Many authors prefer TENAX as an adsorbent because of its low retention of water vapour. Ethyl esters of fatty acids were recovered from *Pseudomonas fragi* cultured on whey, and the stripped volatiles were recovered from TENAX by thermodesorption (Morin et al., 1994). Switching of absorbent columns, as described above, is required in a continuous bioprocess, but the isolation still has to proceed batchwise. Larger-scale processes would require expensive rotary adsorbers and complex instrumentation and control engineering. High performance adsorbents are as valuable as gold. Condensation of highly volatile aroma compounds is described in literature (for example, Dale et al., 1985); however, low temperatures and gas flow rates render open systems expensive. Inherent disadvantages are:

- the high excess of inert gas that lowers the partial pressure of the condensable component,
- the concentration of condensable component is decreased adjacent to the condensate/gas interface, and the driving temperature gradient decreases during condensation,
- the large amount of inert gas must be cooled down, and
- water is usually the major condensable component.

Comparative calculations and experiments were carried out using the benzaldehyde producing fungi *Poria xantha* and *Ischnoderma benzoinum* (Berger et al., 1990). Natural benzaldehyde occurs in the seeds of bitter almond and other stone fruits, and is produced from processing wastes upon hydrolysis of the cyanogenic precursors by distillation. Several reasons have induced industrial interest in the biotechnology of benzaldehyde:

- Hydrolytic generation from plant seeds yields equimolar amounts of hydrogen cyanide (yet $2.5/g \times L^{-1}$ bitter almonds)
- Cherry flavored beverages require large amounts of natural benzaldehyde.
- Chemosynthesis depends on chlorine chemistry and yields nature-identical material.
- Feeding L-phenylalanine to fungal cultures has proven an efficient means to increase product yields (Fig. 10.8; Krahn, 1986; Berger et al., 1990).

Assuming a steady state concentration of 150 mg benzaldehyde in one m^3 of waste air, the molar ratios of the gas phase components can be calculated (Table 10.7). Obviously, extremly low condensation temperatures would be required, and blocking of the condensor cross-section by water ice is bound to occur during the process.

Absorption of benzaldehyde in water using a bubble column as a gas scrubber and a liquid/liquid extractor to regenerate the scrubber permitted continuous operation of the process (Fig. 10.9). No vaccum or distillation facilities and controls were required, except a thermostatted jacket for the recycle of the extraction solvent, *n*-pentane. A charcoal trap was mounted at the outlet of the first column to provide additional safety in case of a failure of the condenser. When making the

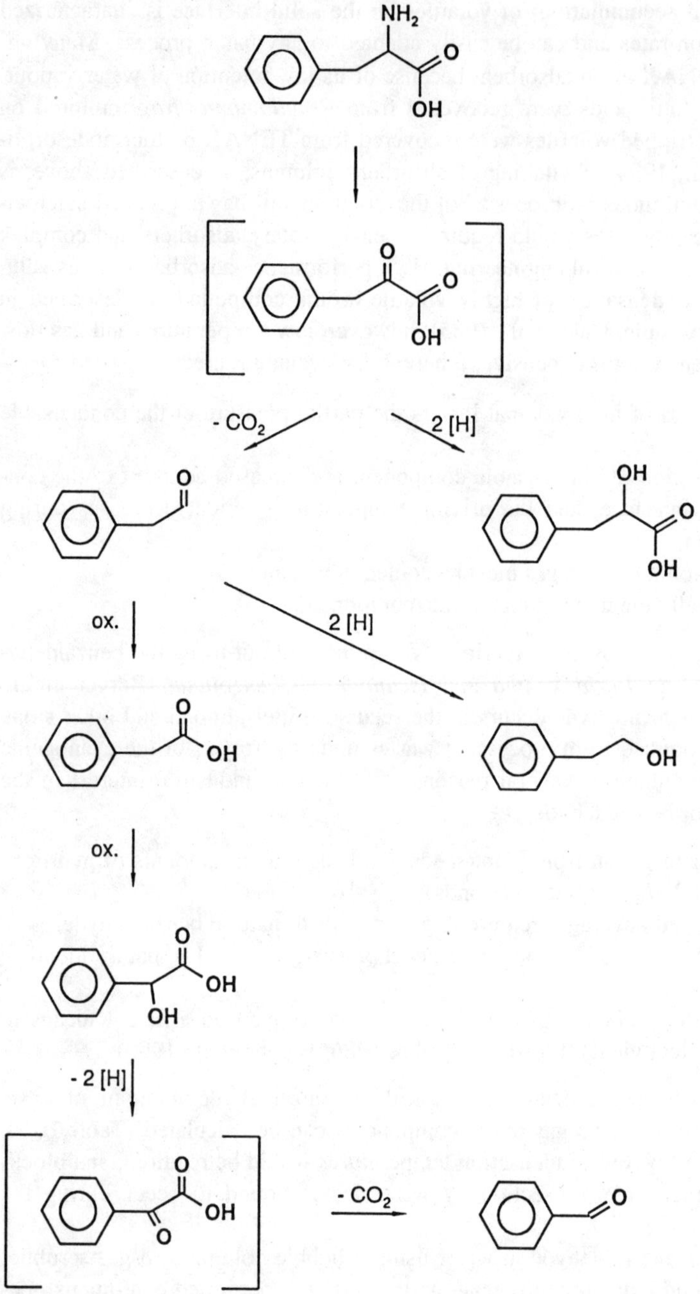

Fig. 10.8. Hypothetic bioconversion of L-phenylalanine to benzaldehyde in cultures of basidiomycetes (modified, after Krahn, 1986)

Table 10.7. Waste air components of submerged cultured *Ischnoderma benzoinum* (Schlebusch, 1992).

Component	Concentration [g m^{-3}]	Molar ratio
Air	1177	97.5978
H$_2$O	18	2.3988
Benzaldehyde	0.15	0.0034

selection of the extraction solvent, boiling point, partition coefficient, and solubility in the absorbing phase should be considered. The basic elements of the process are displayed in Fig. 10.10. About two thirds of the total benzaldehyde formed was recovered by absorptive downstreaming under regular processing conditions of 27 °C and 1.29 vvm. The economics of absorptive downstreaming units are strictly correlated with an efficient carry out of the target compound from the bioreactor.

10.3.9 Membranes

Membranes are key elements in sterile filtration, protect electrodes and sensors, and may, as hollow fiber or flat membrane modules, constitute the bioreactor itself. They are also suited for the separation of sensitive bioproducts (cf. Sect. 10.3.3), because actual isolation is entirely physical and proceeds at ambient temperature. Typical membrane processes applied for downstreaming depend on electrical potential difference (cf. Sect. 10.3.4), on hydrostatic pressure (reverse osmosis), or on vapor pressure or concentration gradients (pervaporation, perstraction).

As has been stated in previous sections, ethanol formation by *Saccharomyces cerevisiae* and butanol/acetone formation by *Clostridium acetobutylicum* are suitable model processes for aroma applications. Pervaporation of ethanol through hydrophobic membranes is being studied under upscaling aspects for the industrial production of low alcohol wines (Fischer, 1994). Operating conditions effected mass fluxes rather than selectivity of the polydimethylsiloxane membrane when synthetic ethanolic and aqueous solutions were separated from volatile aromas (Beaumelle et al., 1992). Mass transfer of esters through the membrane was higher than that of alcohols. The permeation of butyl *n*-butanoate through the same polymer material increased with increasing Reynolds number indicating formation of a polarized layer (concentration polarization) in the course of the mass transfer (Bengtsson et al., 1993). This adverse effect was controlled by optimization of the circulation velocity of the feed. Because pore size distribution cannot be changed much without losing selectivity, total flux has to be increased by increasing the active area of the membrane. Hollow fiber modules have been developed as a space saving unit. Cell recycle by microfiltration and the hollow fiber based sepa-

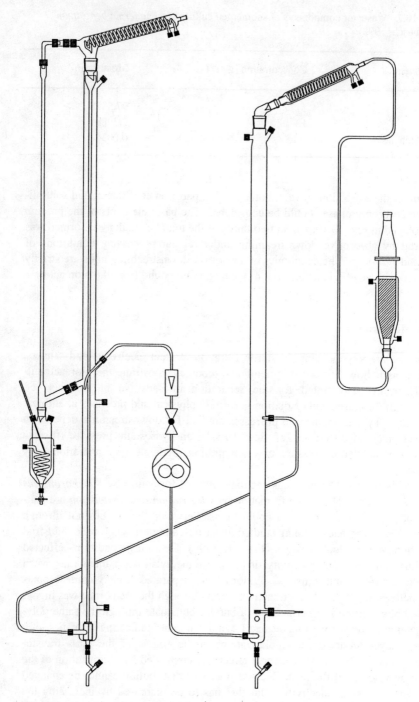

Fig. 10.9. Lab-scale gas absorber coupled to a continuous liquid/liquid extractor (Schlebusch, 1992)

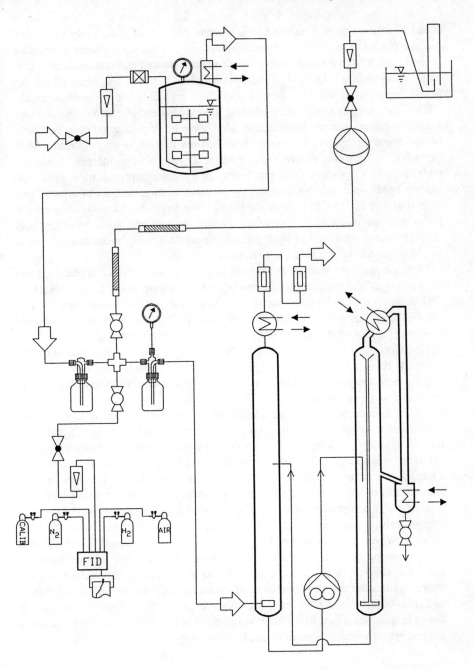

Fig. 10.10. Gas stripping and absorption/extraction for the continuous recovery of benzaldehyde (Schlebusch, 1992)

ration of ethanol were combined to a process that almost completely converted glucose (Groot et al., 1992a). Selectivity and cost of membrane were derived as critical factors from a study comparing membrane assisted and conventional continuous cultivation (Groot et al., 1993). A perstraction system was operated for the in situ removal of volatiles from *Clostridium acetobutylicum* (Grobben et al., 1993). The products were separated using a polypropylene membrane and an oleyl alcohol/*n*-decane mixture as extractant. Membrane fouling occurred upon two days of operation. In a second system, these authors combined microfiltration with extraction by fatty acid methyl esters through a hydrophilic membrane. A comparable increase of product yield was found. In a similar approach membranes enclosing lipids were used to simulate the bioconcentration process of aquatic animals (Lebo et al., 1992). Though the method was suggested to monitor lipophilic pollutants, the majority of aroma compounds possess the same polarities and would be transferred into the lipid. Dialytic or distillative separation from the lipid would be required to obtain a pure aroma compound.

More comparative studies would help us to judge the range of application and the validity of the various membranes in downstreaming volatile bioproducts. A French group has used a microporous membrane and a silicone membrane to recover the fungal impact compound 1-octen-3-ol (Voilley et al., 1990). Few studies have compared membrane processes with other recovery methods (Groot et al., 1992b; Qureshi et al., 1992). Relief of the biocatalyst from product inhibition has regularly been achieved, but design parameters and operational costs differed significantly. Despite their obvious advantages, in situ separation units are used in smaller scale processes only.

Flavor manufacturers have all accumulated remarkable expertise in separating their volatile target compounds from complex natural sources. No fundamental reasons exist for not submitting a microbial nutrient medium of a batch process to the same technologies. The use of pre-installed equipment and machinery will contribute to lower overall costs. The alternative of the future, however, will be continuous processing in a smaller, less expensive bioreactor combined with continuous and in situ product recovery. The successful application of membranes, gas stripping/absorption, liquid/liquid and solid phase extraction has signalling character; however, even the best techniques come along with inherent drawbacks that call for careful consideration: membranes require narrow conditions of operation and act as solid phase extractants; gas stripping is feasible for very volatile aromas only, organic solvents raise questions of toxicity, safety and may form emulsions, and solid adsorbents exhibit low selectivity and cannot be applied to polar constituents in aqueous solution. All achievements cannot hide the rudimentary stage of the present technical designs in aroma biotechnology.

11 Toward an Industrial Application

The ultimate goal of process development is to increase the specific productivity (mass of product per mass of biocatalyst) by optimizing physiological (including genetic) and technological parameters. This sounds fairly straightforward, but experimental work has shown that neither the nutrient composition nor the biochemistry of the cell are as reproducible as a merely chemical process. Among food grade organisms, strains of *Acetobacter* are particularly well known for genetic instability. Subculturing is bound to generate mutants, some of which even may have lost the capability to oxidize ethanol.

Accumulation of intermediates reflects a metabolic situation far removed from homeostasis. The response of single intermediates to precursor and effector concentrations is now expressed by 'sensitivity' and 'elasticity coefficients' that are a measure for the modulation of the respective pathway components. The application of modern theory of metabolic flux to the generation and accumulation of aroma compounds is strongly impeded by the current lack of detailed biochemical knowledge. The general concept of overproduction, however, does not deviate from any other bioprocess: unbalanced growth by nutrient limitation or other unfavorable environmental conditions. For more complex aroma chemicals, additional factors comprise abolition of regulatory mechanisms of 'secondary' pathways (cf. Sect. 5.6), induction of morphological differentiation (cf. Sect. 9.3.2), and pathophysiological permeability of compartments (cf. Sect. 9.3.4). The key roles in nutrient depletion are played by carbon, nitrogen, and phosphorus. Obviously, a lack of either one of these is closely related to cellular growth. Classical work on the regulation of the biosynthesis of β-lactams, acylogenins, and ergot alkaloids can be transferred to aroma targets, if their 'secondary' character has been established. Bioregulators, such as the phytoeffectors (cf. Sect. 9.3.4), nutritional (enzyme) inducers, and precursors are generally applicable means to improve productivity.

11.1 Rapid Process Improvement

Once an interesting volatile has been identified in a lab scale culture, 'optimization' is usually started by repeating the first experiment on different nutrient media. For example the lactic acid bacteria involved in malolactic fermentation of wines (cf. Sect. 2.3.4) are nutritionally demanding and can be differentiated on specific media only (Zuniga et al., 1993). Selective nutrient media are more and more employed to evaluate mixed cultures in the dairy field (Bille et al., 1992). Factorial experiments and statistical treatment of the results have been conducted in order to reduce the number of experiments detecting productivity maxima. Computational strategies, such as simulated annealing or threshold accepting were successfully applied to optimize various industrial processes. Human intuition is not suitable for the optimization of problems (Dueck, 1993). More recently, neural networks were used to design nutrient media (Kennedy et al., 1992). The production of intracellular oil by *Rhodutorula gracilis* was related to three different concentrations of molasses, ammonium nitrate, and yeast extract. Instead of a complete set of 27 experiments, only 10 experiments were needed to predict further test results. Novel biochemical methods, such as DNA probes, DNA affinity columns, and the polymerase chain reaction can assist in the detection and characterization of aroma producing cells (Tannock et al., 1992; Andersen and Omiecinski, 1992). The same purpose is provided by electrophoretic mobility measurements that depend on the electrochemical properties of the cell wall, as outlined above (cf. Sect. 2.1, Bowen et al., 1992). A novel approach for detecting the accumulating compartment in a living cell is in situ staining of enzymes. An example was the localization and subsequent purification of geraniol dehydrogenase from lemon grass leaves (Sangwan et al., 1993). A rapid method for the quantification of the esterase activity of yeast was related to the sensory quality of fermented beverages (Bardi et al., 1993). In situ sensoring will also contribute to a better insight into running a bioprocess and, thus, to a more rapid and concerted improvement. The development of an 'artificial nose' is being currently approached along different routes; the active sensor is either an advanced semiconductor (tin dioxide, tungsten or zinc oxide, phthalocyanines that change the electrical conductivity upon adsorption of a charge carrier) or a biological molecule. As an example of the latter, fractionated olfactory cell proteins of the bullfrog were coated on a piezoelectric quartz crystal connected to an interfaced personal computer. Arranged as a five channel system, the recorded profiles allowed the discrimination of various alcoholic beverages (Wang, 1993). Further research will attempt to combine the long-term stability of the electronic receptors with the selectivity of biological structures. In vivo phosphorus-31 nuclear magnetic resonance was applied to study yeast biochemistry during the formation of ethanol (Taipa et al., 1993). Distinct profiles for the variation of both intracellular and extracellular pH were found. Immobilized cells maintained internal and external pH better than the freely suspended cells indicating a regulatory effect of the gel matrix (cf. Sects. 2.2, 2.3.5). This in turn was concluded to favor the formation of the volatile product.

The optimization and control of bioprocesses is more and more achieved by computer based handling of fermentation parameters. A learning system that required no modelling was used to determine the optimum dilution rate of a continuous stirred tank reactor (Thibault and Najim, 1993). For the optimization of a fedbatch process iterative dynamic programming was reported (Luus, 1993). Fuzzy logic control has already made its way into biotechnology (Table 11.1). The parameters of a fuzzy model can be selected by the classical simplex method and the

Table 11.1. Fuzzy logic in biotechnology

Process	Reference
Batch culture of *Zymomonas mobilis*	Jitsufuchi et al., 1992
Fed-batch for penicillin production	Nyttle and Chidambaram, 1993
Fed-batch of *Saccharomyces cerevisiae*	Park et al., 1993
Large scale beer brewing	Simutis et al., 1993

operator's knowledge. A fuzzy expert system processed the output of several artificial neural nets that were used as a set to control parts of the entire process (Simutis et al., 1993). The process knowledge of the nets was continuously updated by a real-time learning function upon activation. The reader is referred to a recent textbook for details of application of fuzzy sets (Zimmermann, 1991).

11.1.1 Improved Biotransformation

Due to the frequent need to use arbitrary standards and the possible occurrence of side optima of productivity 'optimization' is not a stringent term in biotechnology. Many authors, however, use the term to indicate significant improvements of an existing process. General options to improve a biotransformation characterized by a lipophilic substrate chemistry were compiled in Table 11.2. Usually, more than one aspect has to be dealt with to achieve an 'optimized' process.

Adapted cells of *Pseudomonas putida,* an often used catalyst in biooxygenation reactions (cf. Sect. 6.1), were obtained by precultivation on (+)-camphor (Grogan et al., 1993). Monooxygenases were induced and oxygenated various ketone substrates stereoselectively. A mixed culture of anaerobic bacteria was pregrown on a

Table 11.2. Options to improve the transformation of lipopholic substrates

• Improved biology:	selection of strain, adapted cells, enrichment culture, modified enzyme techniques
• Improved chemistry:	nutrient limitation, continuous addition of nontoxic amounts of substrate in nontoxic solvents (glycols ets.), substrate derivatives, selective inhibitors, competing carbon source, additives
• Improved processing	low water systems, cofactor recycle, *on line* monitoring, immobilization
• Improved downstreaming	integrated systems, *in situ* recovery

dichloromethane-containing medium for dehalogenation processes (Braus-Stro-meyer et al., 1993). For monochloro methane a highly efficient transalkylation was obtained. The enzymatic type of the reaction was experimentally demonstrated. Single step reactions are now also approached using modified enzymes. Tailor-made biocatalysts were prepared using a wide variety of chemosynthetic tools. A lipase of *Pseudomonas cepacia* was chemically coupled to a polyethylene glycol derivative to render the enzyme soluble in organic solvents (Kodera et al., 1993). The activity was maintained in the modified form, as demonstrated by the trans-formation of 16-hydroxyhexadecanoic acid to the corresponding macrocyclic lac-tone. 15 to 17-membered cyclic ketones are impact compounds of extremely valu-able animal products, such as musk and civet. Macrolides and macrocyclic dilac-tones are widely used as musk imitators in industrial flavorings. The chemosyn-thesis of 16-hexadecanolide is possible starting from tricyclohexylidene peroxide, but has been cancelled as early as the 1970s for safety reasons.

Derived from general biotechnological experience, attempts to improve the chemical environment for the transformation of volatile compounds by nutrient limitation have become known (Targonski, 1992): Methoxy benzaldehydes, such as 3,4-dimethoxy benzaldehyde with its vanilla-like odor were transformed by aerobic cultures of *Pichia stipidis*; nitrogen depleted cultures were found to be more active. The large scale preparation of the chiral synthon (*S*)-ethyl 3-hydroxybutanoate by baker's yeast mediated reduction of ethyl acetoacetate was improved in terms of absolute yield and optical purity, when carbohydrate was substituted by ethanol as the sole source of energy (Kometani et al., 1993). A fed-batch operated bubble column reactor was suggested for a technical realization. Cytotoxic and lipophilic substrates have to be supplied continuously to active cultures (cf. chapter 6). Derivatives of the transformation substrates can contribute to overcoming these problems, but with increased polarity the molecules diffuse less readily through the cell membranes. Further drawbacks associated with substrate conjugates are natural occurrence, and the possible degradation by cellu-lar activities before reaching the target site. However, some success was achieved in steroid biotransformations.

Many biotransformations yield a whole spectrum of side products in addition to the desired structure. One measure is to perform the reaction in the presence of a selective inhibitor. The unusual (*S*)-isomer of 3-hydroxypentanoate was obtained from ethyl 3-oxopentanoate in a baker's yeast catalyzed reduction upon addition of alkyl bromide (Ushio et al., 1993). (*R*)-enzyme activity was strongly suppressed. As the inhibitory effect depended on the time of preincubation, a covalent modifi-cation of the (*R*)-enzyme was presumed. Though the (*S*)-enzyme was also slightly subject to inhibition the method proved useful in the preparation of chiral 3-hydroxy esters in high enantiomeric excesses. Some of these esters contribute to fruity notes of exotic fruits. Biotransformations prone to the formation of side products can also be improved by co-feeding of competing carbon substrates, because the cell will shift to the better metabolizable substrate and refrain from attacking the product molecule further. Bipolar compounds were used as additives to improve the water solubility of the substrate or to replace thermal substrate activation (Majcherczyk et al., 1993). Polyvinylpyrrolidone and polyoxyethylated vegetable oil increased the yields in hydroxylation reactions of substrate with low water solubility (Chien and Rosazza, 1980). Lecithin reversed the inhibition of chitin synthase (Pfefferle et al., 1990). Information on such activators or protectants appears scattered in the literature, and most of the positive findings were not the result of systematic screenings.

The successful application of enzymes in microaqueous systems for the generation of flavors (cf. chapter 7) and other food ingredients (Vulfson, 1993) has given reasons to cultivate whole cells under similar conditions. The hydrophobicity of a solvent was suggested as a measure of cytotoxicity (Vermue et al.,1993). The pyruvate decarboxylase catalyzed acyloin formation from benzaldehyde and pyruvate was investigated using intact *Saccharomyces cerevisiae* in organic solvents (Nikolova and Ward, 1992). Maximum biotransformation activity was observed in a hexane/water (9/1) two phase system. Lowest activities occurred with toluene and trichloro methane. Scanning electron microscopy revealed damage in the form of cell puncturing after different exposure of the yeast to trichloro methane, butyl acetate, and ethyl acetate. In nutrient media that contained damaged cells increased amounts of phospholipids were detected. It appeared that more polar solvents extracted bipolar, structurally essential constituents of the outer membrane. This work has significantly contributed to our understanding of the overlapping events of increased permeabilization and loss of biocatalytic activity (cf. Sect. 9.3.4). In an even more drastic experiment cells of *Arthrobacter simplex* were positioned between an agar layer and an hydrophobic solvent (Oda and Ohta, 1992). Hydrophobic substrates were biotransformed more efficiently than in emulsified systems. The approach was termed 'Interface Bioreactor'. A strain of *Pseudomonas putida* was adapted to growth under a continuous layer of *p*-xylene (Wolfram et al., 1992). Total catabolism rather than partial conversion was the aim of this study. Lack of cofactor recycle as a general drawback of enzyme based redox reactions has been mentioned above (cf. Sect. 7.6). Alternative coupled enzyme systems using alcohol dehydrogenase or formate dehydrogenase in monooxygenase catalyzed Baeyer-Villiger oxidations for the in situ recycling of $NAD^+/NADH$ are currently being developed (Grogan et al., 1992). The group of Simon has investigated numerous high yielding redox reactions of smaller molecules (for example, Schinschel and Simon, 1993). A chemical redox mediator is used in simulating biological electron chains. The mediator compound was regenerated by either enzyme coupling or electrochemically. Space/time yields in the range of $kg \times L^{-1} \, d^{-1}$ were reported. The legal status of such extremely promising systems needs to be evaluated.

Though the initial enthusiasm has been replaced by a more realistic view, immobilization of enzymes and whole cells is still regarded as an important tool of process improvement. Various studies on the biotransformation of steroids have model character for aroma biotechnology. A regioselective dehydrogenation by *Arthrobacter simplex* performed best by immobilizing the cells in a polyurethane foam (Pinheiro and Cabral, 1992). The liquid reaction phase was *n*-decanol. Cholesterol oxidation was studied in liposomal media consisting of phosphatidylcholine vesicles (Goetschel et al., 1992). In the presence of cells of *Rhodococcus* the natural bipolar compound was not metabolically inert. Strains of *Candida tropicalis* and *Trichosporon cutaneum* were covalently linked to cellulose granules (Ivanova and Yotova, 1993). The immobilized cells were used to study the biotransformation of furfural. Horse liver alcohol dehydrogenase was entrapped in a silicone polymer to operate the catalyst in a packed-bed (Kawakami et al., 1992). The enzymatic oxidation of alcohols or the reduction of aldehydes with cofactor recycle was demonstrate in *n*-hexane. The same authors modified the hydrophobicity of the support by adding Ca alginate for the transformation of alkenes into epoxides by *Nocardia corallina*.

These examples suggest that observation of specific properties of aroma compounds and the according modification of existing processes hold much promise for improved performance. Exotic measures, however, such as using argon gas as a protectant in the bioconversion of volatile phenols by yeast (Huang et al., 1993), will remain rare exceptions.

11.1.2 Improved De Novo Synthesis

Most of the rules for the transformation of lipophilic compounds also apply to de novo synthesis (Table 11.2). When dealing with more complex eukaryotics not only strain selection and adaptation, but other biological parameters such as inoculum preparation and size, spore formation, and cell aggregation may affect productivity (Abraham et al., 1994). Similar observations were made with the methyl anthranilate producing fungus *Pycnoporus cinnabarinus* (Gross et al., 1990). This study has also demonstrated the nutritional sensitivity of higher fungi: Two different pepsic meat peptones oppositely influenced the formation of volatiles. While a maximum of methyl anthranilate was formed in the presence of maltose, no aroma formation occurred in the malt extract medium, though this carbon source contained >80% maltose. Actual mechanisms of this carbohydrate regulation of aroma metabolism have not yet been elucidated. As a result, it is not surprising that random nutritional variation (Hanssen et al., 1984) and factorial experimental designs (Christen and Raimbault, 1991) were applied to locate optima of aroma formation. Even Omelianski's ideas (cf. Sect. 5.1) were revived in a patent describing the formation of cheese aroma by *Monascus purpureus* in a milk powder supplemented nutrient medium (Rütgerswerke AG, 1992).

A novel technique has been described to prevent cytotoxic levels of lipophilic substrates: A silicon tubing was placed in the reactor, and the liquid substrate was circulated within the tubing from a reservoir (Choi et al., 1992). The transfer of substrate was mainly effected by the speed of the impeller indicating critical mass transfer at the solid/liquid interface. Again the system was intended to biodegrade waste solvents, such as toluene or xylenes, but the approach appears immediately transferable to the bioconversion of monoterpenes and other aroma compounds. 'Natural' downstreaming of volatiles using adsorbents (cf. Sects. 10.3.7, 10.3.8) has gained much popularity since the first successful reports (Schindler, 1982).

Modern bioprocesses have used adsorbent trapping as an integrated part of the system (van der Schaft et al., 1992).

In conclusion, a number of key parameters have been identified that have led to rapid improvement of novel lab scale aroma generation bioprocesses. However, neither the chemical environment nor fundamental physical parameters, such as illumination (Delgado et al., 1992), have been researched to an extent that would permit general recommendations for a systematic and concerted working scheme.

11.2 Scaling-Up

Character impact components of natural aromas (cf. Fig. 1.1) determine the overall flavor of a food item in trace concentrations. Therefore, an annual production of some hundred grams may meet the market demand. In this respect scaling-up will, for a great number of target compounds, remain a minor problem of aroma biotechnology. Other bioflavors, such as carboxylic acid esters will be required in larger quantities (cf. chapter 5).

A recent review has addressed major problems of scaling-up and included actual case studies and checklists for project planning (Reisman, 1993). The step from flask culture to laboratory scale bioreactors often resulted in process improvements, because key parameters, such as pH and pO_2 were better controllable in the reactor system (Burla et al., 1992). However, even in medium sized tank reactors, heat and mass gradients appear inevitable. The resulting metabolic heterogeneity of the cell population is reflected by macroscopic fluctuations. A fed-batch culture of *Saccharomyces cerevisiae*, for example, could not be synchronized due to inherent glucose gradients in a stirred tank reactor (Namdev et al., 1992). Emphasizing the microenvironment of the producing cell the principle of physiological similarity was proposed as a guideline for up-scaling (Votruba and Sobotka, 1992). As physiological similarity ultimately depends on chemical, thermal, and fluiddynamic parameters, classical similarity theory must be applied to scale-up those parameters that exert the most significant physiological effects. Unfortunately, these key parameters are difficult to evaluate for certain reactor types that possess particular potential in aroma biotechnology. For example, this holds true for solid state reactors (Durand et al., 1993).

11.3 Profitability Aspects

The increasing industrial application of aroma biotechnology is documented by the increasing number of patents (cf. chapter 5) and marketable products (cf. Sect. 11.4). As the example of vanillin showed (cf. Sect. 5.4.1), differences in price between the 'natural' compound and its synthetic counterpart of one to two orders of magnitude are sufficient reasons. An additional impetus is the increasing proportion of natural aromas in flavoring important groups of food products

Table 11.3. Proportion of natural aromas (%) in flavored food (Hariel, 1993)

Product group	Europe	US
Nonalcoholic beverages	90	83
Dairy products	50	75
Savoury products	80	80

(Table 11.3). In 1990, an estimated 300 metric tons of aroma compounds were produced biotechnologically; enough to flavor several millions of tons of food items.

In chosing target compounds, the potential sales, the unit selling price, and the stability of demand will have to be assessed in order to recoup the research and development costs. As for all chemical processes, the major cost factors are substrate cost, cost of further chemicals, solvents, labor, energy, taxes, and durable equipment. In chosing the most suitable biocatalyst, immobilized enzymes should be preferred wherever possible. Enzymes benefit most from stabilization by immobilization, and can be easily retained in the process by using membranes. No substrate has to be wasted for supporting all other reactions that are not involved in product formation. Intact cells provide self-regeneration of catalytic potential and of cofactors, active transport systems, and evolutionary optimized enzyme chains. Strains of *Schizophyta* and *Protoascomycotina* used in or related to classical food bioprocesses are good starting points. Higher fungi and plant cells exceed the potential of less organized cells, but require skillful handling.

The production cost of a bioflavor is determined by the conversion yield of the respective target compound, but further factors are contributory:

- Inexpensive constituents of the nutrient medium; side streams of meat, fish, fruit, and vegetables processing and from the dairy and fermentation industry can be used.
- High space/time throughput of the substrate will save energy cost and capital investment.
- Minimized cost to avoid environmental pollution; risks can be seen in the use of xenobiotic additives, such as 2,4-dichlorophenoxy acetic acid in plant cell culture, or in the accumulation of toxins in the spent biomass.
- Efficient downstreaming. Batch distillation or extraction implicate the handling of bulk amounts of water and should fall into disuse. More selective techniques based on the chemical differences of the target compounds and bulk constituents have become available (cf. chapter 10).

Very few data on actual processing cost have been published. A simple yeast mediated biotransformation with 100% conversion within one day may deliver a product at 110% of the substrate cost (invested capital and depreciation excluded).

A current average market price of bioflavors of 150 to 200 US $\times$kg^{-1} has been mentioned (Armstrong at al., 1989); however, some of the first commercial products were priced at > 10 000 US $\times$kg^{-1}. A heavy price tag may be just the result of low yields and productivities, but also of the cost of the substrate that has to be classified 'natural'. Derived from published data on lab scale bioprocesses, a 50% conversion rate within 48 h to yield 500 mg per liter of product can be assumed as a good starting point. Further assumptions (labor cost and downstreaming 300 $ per m^3 and day, 20 $ per m^3 and day nutrient cost) lead to a cost price of 1480 $ per kg, if the natural substrate was available at 100 $\times$kg^{-1}; a substrate at 1000 $\times$kg^{-1} would result in cost of 3280 $\times$kg^{-1}. The prominent role of the substrate cost has been recently underscored by model calculations (Armstrong, 1993): Unit cost to produce benzaldehyde (cf. Sects. 5.4, 10.3.8) from glucose were about 4 $ per kg, while the same product obtained by enzymatic oxidation of benzyl alcohol would cost about 750 $ per kg. This justifies the large number of attempts to synthesize volatile flavors de novo (cf. chapter 5). The same study also detailed the cost of equipment for a stirred tank process (1 m^3 working volume) and a comparable enzyme column reactor. Total cost was in a similar order of magnitude for both variants and amounted to 600 to 700 $\times$kg^{-1}. The selling price of a volatile phytochemical should be in the range of 2000 to 5000 $\times$kg^{-1} to render plant cell culture economic. High expectations were tied to continuously operated processes using immobilized plant cells. Production cost in the range of 20 to 25 $\times$kg^{-1} were predicted (Moshy, 1985). So far, no experimental system has reached a productivity to match these predictions.

Looking over these calculations and estimations it appears that a number of industrial processes obviously work out very well. Other processes with low productivities or expensive substrates are operated to yield volatile compounds retained for captive use in the company's own flavor blends. Some of the larger aroma producers have reduced their investment cost by giving R & D contracts to small venture capital companies or university institutes to test if aroma biotechnology is applicable to their product needs. If a project turns out promising, in-house research is linked-up to bringing cost down to meet the market.

11.4 Present Industrial Applications

Since a few years companies have started to advertise '100% natural' aroma chemicals. If these were obtained on classical routes or if they, in fact, originated from a bioprocess is usually difficult to decide. Common volatiles of fermentative pathways can be enriched in concerted processes by appropriate measures. Among these volatiles are trivial, but nevertheless important chemicals, such as ethanol, acetone, n-butanol, fermentation acids, diacetyl, and the different types of cheese, soy, and yeast flavors (cf. chapter 2). Distilled or extracted side streams of empirical biotechnologics were used as a source of aroma. Examples are starter culture distillates (diacetyl) and lie de vin oil ('cognac oil', wine lees oil') from steam distilled deposits formed during wine making.

The first generation of aroma compounds from biotechnology arose with the availability of lipases active in microaqueous environment (cf. Sect. 7.1.3). A whole series of fusel oil constituents were able to be transferred to acyl moieties of natural plant oils using biocatalysts. Methyl branched fatty acids were prepared from the corresponding alcohols, and aromatic acids, such as benzoic, cinnamic, anthranilic, and salicylic acid were isolated from essential oils for further esterification. Similarly, the initially small range of alcohols was expanded by microbial and plant metabolites such as (Z3)-hexenol, phenyl ethanol, cinnamyl alcohol, and various terpenols. Theoretically, about 400 carboxylic esters are amenable to the biotechnological approach, which is about one third of all esters identified. However, a considerable number of the theoretical ester products is without any sensory significance. Natural aldehydes may be obtained from the corresponding molecules by biocatalyzed redox reactions or STRECKER-type degradation, from the lipoxygenase pathway (cf.Sect. 9.1.1), or from food processing (acetaldehyde from wine fractions, benzaldehyde from bitter almond oil). Natural methyl ketones can be produced according to their formation in blue cheese by degrading pure, even-numbered fatty acid precursors. Natural 4- and 5-alkanolides have been claimed in numerous patents to be derived from precursor hydroxy acids of natural origins (cf. Sect. 5.4.3). The lactones are valuable volatiles in apricot, coconut, dairy, peach, and other fruit flavors. As the biochemistry of fatty acids is generally even-numbered, natural 4-undecanolide is a remarkable product. Only a few other heterocycles are available on the market. 2,5-dimethyl-4-hydroxy-3(2H)-furan-3-one, useful in strawberry, pineapple, and thermal aromas, is manufactured in a multistep bioprocess from L-rhamnose (cf. Sect. 7.2.2).

Norisoprenoids with ionone, irone, and damascone structure are important modifiers in fruit aromas (cf. Fig. 7.1). In the intact plant many norisoprenoids occur predominantly in bound forms (cf. Sect. 7.2.1) which offers potential for the use of glycosidases and other hydrolases. Thus, the increase of irones in 'fermented' rhizomes of *Iris pallida* is most likely due to hydrolytic activities. The classical process includes maturing, grinding, and steam distillation, and, consequently, favors liberation of the key compounds on each single step. The product, 'orris root oil', belongs to the most precious natural aromas (Fig. 11.1). A luring route to norisoprenoids and related trimethyl cyclohexanes is the oxidative degradation of

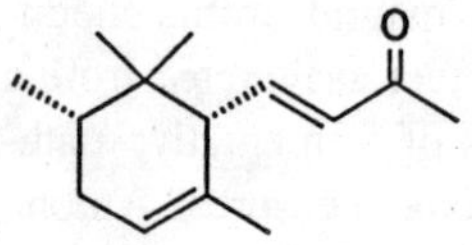

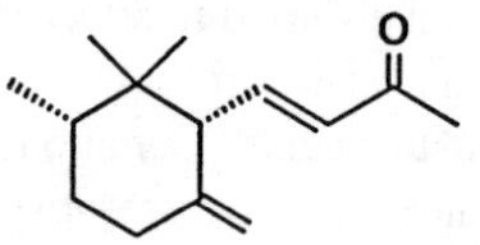

Fig. 11.1. Impacts of orris root oil. I = *cis*-α-irone, II = *cis*-γ-irone

natural precursor carotenoids. The photooxidation of β-carotine led to the identification of 27 volatiles, mostly with distinct sensory properties (Hohler et al., 1988). Acyclic tetraterpenoid isolates would result in linear, methyl branched volatile flavors (Fig. 11.2). Various methods of oxidation are possible:

- Chemically using peroxides, etc.,
- chemically using photooxidation or other $^1O_2$-generating systems,
- chemically via radicals using nonsensibilisized photoreactions, or
- enzymatically using peroxides generated by lipoxygenase activity.

Microbial systems have not yet been published. A recent patent described multiphase mixtures of soy flour as a source of lipoxygenase/hydroperoxide lyase, extracts of plants (carrot, *Carthamus*), and oxygen for the generation of α- and β-ionone, dihydroactinidiolide, *n*-hexanal, and 2,4-decadienals (BFA Laboratories, 1994). A number of ionone mixtures are offered at the market place, but, as natural food constituents would serve as precursors in all cases, it is analytically difficult to draw conclusions on the process used. After all, isolation from *Osmanthus* oil or other plant sources can, at least theoretically, not be ruled out. A commercial product is said to be derived from an aldolase catalyzed process starting with natural citral and acetone, the equivalent of the chemosynthesis according to TIEMANN. This and further processes, such as acid catalyzed esterification, retroaldol condensations for the generation of aldehydes, pyrolysis of non-food materials for

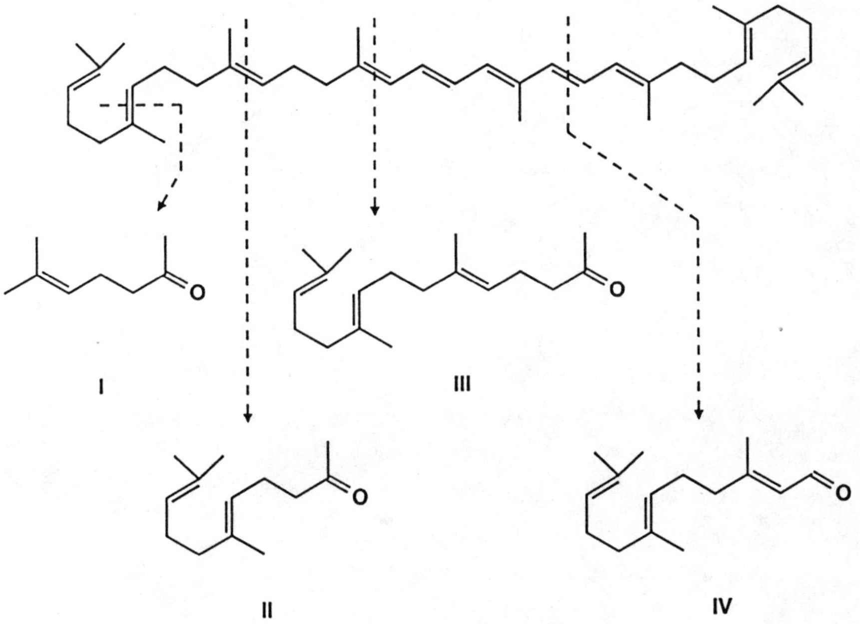

Fig. 11.2. Primary cleavage products from a tetraterpene. I = 6-Methyl-5-hepten-2-one, II = Geranylacetone, III = Farnesyl acetone, IV = Farnesal

furanoid compounds and maltol, and intentional degradation of eugenol or glycerol by 'natural' chemicals belong to what has been called 'soft chemistry'. Products originating from these processes have been accepted by flavorist in the US, and are being introduced more and more into European products. An integrated evaluation of the analytical data sometimes points to a nature identical status of these chemicals. However, as long as an alternative natural aroma chemical is not available on the market and the producer certifies 'natural' status, the flavorist will not hesitate to round off his blends with a doubtful constituent.

The average consumer discriminates chemosynthetic aroma compounds against natural ones much more clearly than they are discriminated in the chemical reality. Being afraid of all the presumed 'chemistry' in his food, he just feels safer with naturally flavored products. That a natural raspberry aroma extracted from the fruit would have to be priced beyond 100 000 $\times kg^{-1}$, and that the origins of a chemical are irrelevant for the toxicological status is simply ignored. As a result and in contrast to many expectations, the market demand for natural aroma compounds has been steadily growing over the past decades. A multinational company has incorporated 87.5% natural flavors in its products worldwide, 7.2% were nature identical, and 5.3% artificial (Hariel, 1993). The legally defined categories assist in misleading the consumer: 'No compounds are made on earth other than those permitted by the laws of nature' (Teranishi, 1989).

12 Outlook

The history of 'bioflavors' is long (cf. chapter 2, Sect. 5.1) and has, more recently, been successfully continued by bioprocesses for nonvolatile flavors, such as HFCS, MSG, or aspartame. As the above chapters have amply demonstrated, there have been numerous achievements in the field of volatile flavors, and there exist many more potential applications. One critical factor will be the formulation of achievable objectives and target compounds (cf. Sect. 11.3). A more rational development of both biocatalysts and processing techniques is now possible on the basis of accumulated knowledge.

In developed countries the basic human necessities of life are satisfied, and the maintenance of health determines many activities including the selection of food. Healthy lifestyles cast food more and more in the role of pharmaceuticals ('nutraceuticals'), but how to accept a low fat or a meat free product without a fat or meat aroma? 'Clean labeling' aims at banning 'non-natural' food additives; can natural aroma compounds with antioxidative or preservative properties (volatile phenols, terpenoids, cf. chapter 1) be functional substitutes for traditional inhibitors in certain products? Volatile flavors affect mood and general well-being: compounds with relaxing or stimulating activities are incorporated in chewing gum and candies, and again the ingredients should be of 'natural' origin (Otte, 1994): All of these aspects are highly compatible with biotechnologically generated aroma compounds (cf. Sects. 1.3, 3.1). Lipases, other hydrolases, and, to a lesser extent, oxidoreductases are being used in viable processes (cf. chapter 7). Even larger is the hardly explored potential for enzymes that catalyze carbon-carbon bond formation (cf. Sect. 6.1). The profound virtues of enzymes, unsurpassed by most man-made catalysts, can be altered by site-directed mutagenesis to tailor the biocatalyst for a desired product. More and more, the amino acid residues of the catalytic site of enzymes are being identified: Lipases and, more recently, a liver alcohol dehydrogenase are examples of such well-characterized, aroma related enzymes (Roig and Kennedy, 1992). A fascinating concept is the use of catalytic antibodies ('abzymes'). Antibodies to a stable mimic of the transition state of a reaction are raised, and then selected from a mixture according to the smallest K_m for the transformation under investigation. Hydrolyses are a likely field of future applications, while the imitation of dynamic enzymic mechanisms will not be feasible. 'Synzymes', synthetic enzyme analogues or mimics without peptide backbone, hold even greater promise, because they can be adapted to many processes. An alcohol dehydrogenase from baker's yeast was doped with extra binding sites for NAD (Yang and Russell, 1993). When the enzyme was preincubated in heptane, the initial activity was increased threefold.

In a technical process elevated temperatures increase substrate turnover and de-crease the risk of microbial contamination. These unphysiological conditions create stability problems. Chemical encagement into polyacrylamide derivatives resulted in a significant enhancement of the thermal stability of several redox en-zymes (Lehn et al., 1992). K_m-values and pH optima of the stabilized enzymes remained unaltered. Various designer enzyme concepts have been recently revie-wed (Wiseman, 1993). Aspects often cited for future research are the elucidation of molecular events leading to enzyme inactivation, processes in extreme environ-ments and with nonnatural substrates, and enzymes redesigned using genetic engi-neering. Since kits for the expression and purification of recombinant proteins can be ordered by phone, larger quantities of every enzyme are available if the amino acid sequence information is known (cf. Sect. 8.2).

Aroma biotechnology using bacteria, yeasts, higher fungi, or plant cells is im-peded by a lack of biochemical understanding. Only a few pathways have been explored in detail:

- lipoxygenase induced catabolism of unsaturated fatty acids,
- monoterpene synthesis including cyclization reaction and formation of irregular olefinic terpenoids,
- degradation reactions of amino acids,
- formation of thio compounds in *Allium* and *Brassica* species, and
- some aspects of phenyl propane metabolism (Schreier and Winterhalter, 1993, Teranishi et al., 1992).

A major experimental drawback to studying aroma biosynthesis on an enzymatic level are the tiny amounts of enzyme protein in the aroma producing cells and the associated difficulty of isolating them from the complex matrix. Once a protein has been purified, monospecific polyclonal antibodies can be raised and used to screen cDNA expression libraries constructed with mRNA isolated from the enzyme source. The corresponding cDNAs can be isolated and sequenced, and the encoded enzymes expressed in a heterologous host. Implementation of a key catalytic activity into a microbial cell would then lead to the (increased) production of cer-tain, even plant aroma compounds. This application of recombinant DNA methods was suggested for studying monoterpene cyclases (Croteau, 1993). Very few data have been published on gene transfer between the different taxa of classical mixed cultures (cf. Sect. 2.10). Another, more rational approach of combining catalytic activities of different sources is coimmobilization of enzymes and intact cells (Hartmeier, 1990). The formation of ethanol from cellobiose by a coencapsulated yeast/β-glucosidase system indicated the possible feasability of generating other volatile flavors.

In tandem with the development of strain and medium, an efficient mode of operation of the process should be considered. Continuous culture systems are often coupled to cell-recycling units or operated using an immobilized catalyst (cf. Sect. 10.1.1). As the scaling-up of a continuous process can magnify problems of sterility and strain stability, commercial processes often try to simulate continuous mode in batch equipment by using 'repeated batch' concepts or the like. All engineering decisions must be based on physiological data. For example, when a repeated batch cultivation of *Bacillus subtilis* or *Bacillus licheniformis* was performed in a two phase aqueous system (cf. Sect. 10.3.6), the cells were partitioned to the bottom phase, and the *exo*-α-amylase was in the top phase. Starch, the inducer of enzyme formation, and glucose, its repressor, were partitioned accordingly to reach increased productivity during consecutive stages of production (Stredansky et al., 1993).

Metabolic precursors, analogues, or inhibitors can be used to redirect the carbon flow of the cell. Liposomes or electroporation eliminate the problem of placing molecules into cells that would not normally cross the membrane because of their polarity or size. Electroporation worked with food species, such as *Lactobacillus* (Aymerich et al., 1993), *Saccharomyces cerevisiae* (Ganeva and Tsoneva, 1993), *Kluyveromyces lactis* (Russell et al., 1993), but also with plant protoplasts (Hidaka and Omura, 1993). The most common application of electroporation has been the introduction of foreign nucleic acids. Generally, the application of rDNA techniques in aroma biotechnology is not limited to the construction of improved enzymes and strains (cf. chapter 8). Working protocols based on genetic techniques hold promise for all stages of process development. The identification of strains using mDNA restriction analysis has been related to effects of yeast strains on the aroma of wine (Querol et al., 1992). DNA probes served the same purpose (Schleifer, 1990): These probes are single stranded fragments that interact with a complementary sequence of a target nucleic acid. In contrast to immunological methods no interference by environmental conditions is to be expected. Antisense DNA or RNA sequences likewise hybridize with a target sequence that encodes for a certain protein. If this targeted protein is involved in the catabolism of an aroma compound, this degradation would be inhibited to permit product accumulation. The polymerase chain reaction requires short oligonucleotides complementary to sequences on either side of a target site. A polymerase then synthesizes a new complementary strand. By repeating a cycle of denaturing, annealing the primer sequence, and polymerization the target sequence is amplified in vitro millions of times. As the target DNA accumulates faster than bacteria can grow, one envisaged application could be monitoring of recurrent contaminants of a bioprocess (cf. Sect. 10.2).

After decades of speculations, modern molecular techniques are now revealing the biochemical mechanisms of odor perception (Shepherd, 1993). A cDNA for an olfactory marker protein has been cloned and sequenced, and no homology with any other protein was found. The sequence of an odorant binding protein resulted in a tertiary structure similar to proteins that carry small lipophilic molecules (apolipoprotein D/cholesterol; β-lactoglobulin/retinol). The protein formed a cuplike structure with a hydrophobic central cavity and was supposed to function as a carrier and filter of odorants. Olfactory neurons were transferred to in vitro culture and shown to respond to physiological levels of volatiles (Graziadei, 1993; John Hopkins University, 1993). In agreement with previous hypotheses on the involvement of G proteins and cAMP, the stimulation of adenylate cyclase (and phosphoinositide turnover) was experimentally confirmed (Ronnett et al., 1993). A multigene family encoding for mammalian odorant receptors was identified (Buck, 1993). Different genes appear to be arranged in distinct anatomical patterns within the olfactory epithelium. A set of complementary DNA has been patented (Columbia University, 1992), and expressed in non-neuronal cells (Raming et al., 1993). A first practical application has been developed by coupling an odorant binding protein from bovine nasal mucosa to a CNBr-activated resin (Bussolati et al., 1993). This opens new perspectives for the preparation of both sensors and affinity adsorbents with an unsurpassable selectivity for odor molecules. In situ measurement and recovery of even traces of single volatile target compounds will come within reach.

References

Chapter 1

Acree FE, Barnard J, Cunningham DG, A procedure for the sensory analysis of chromatographic effluents, Food Chem, 1984, **14**, 273

Bartlet E, Blight MM, Hick AJ, Williams IH, The responses of the cabbage seed weevil (*Ceutorhynchus assimilis*) to the odour of oilseed rape (*Brassica napus*) and to some volatile isothiocyanates, Entomol Exp Appl, 1993, **68**, 295

Bauer K, Garbe D, Surburg H 1990, Common Fragrance and Flavor Materials, VCH Weinheim

Berger RG, Drawert F, Nitz S, Sesquiterpene hydrocarbons in pineapple fruit, J Agric Food Chem, 1983, **31**, 1237

Bradow JM, Inhibitions of cotton seedling growth by volatile ketones emitted by cover crop residues, J Chem Ecol, 1993, **19**, 1085

Burdach KJ 1988, Geschmack und Geruch, Huber Bern

Byers JA, Attraction of bark beetles, *Tomicus piniperda*, *Hylurgops palliatus*, and *Trypodendron domesticum* and other insects to short-chain alcohols and monoterpenes, J Chem Ecol, 1992, **18**, 2385

Croft KPC, Juttner F, Slusarenko AJ, Volatile Products of the lipoxygenase pathway evolved from *Phaseolus vulgaris* (L) leaves inoculated with *Pseudomonas-syringae* cv-*phaseolicola*, Plant Physiology, 1993, **101**, 13

Cruickshank RH, Wade GC, The activation of latent infections of *Monilinia fructicola* on apricots by volatiles from the ripening fruit, J Phytopathol, 1992, **136**, 107

Deng WL, Hamilton-Kemp TR, Nielsen MT, Andersen RA, Collins GB, Hildebrand DF, Effects of 6-carbon aldehydes and alcohols on bacterial proliferation, J Agric Food Chem, 1993, **41**, 506

Donovan A, Isaac S, Collin HA, Inhibitory effects of essential oil components extracted from celery (*Apium graveolens*) on the growth of *Septoria apiicola*, causal agent of leaf spot disease, Plant Pathology, 1993, **42**, 691

Dwumfour EF, Volatile substances evoking orientation in the predatory flowerbug *Anthocoris nemorum* (Heteroptera, Anthocoridae), Bull Entomol Res, 1992, **82**, 465

Elizalde BE, Bressa F, Rosa MD, Antioxidative action of maillard reaction volatiles - Influence of maillard solution browning level , JAOCS, 1992, **69**, 331

Emberger R, An analytical approach to flavor research, Cereal Foods World 1985, **30**, 691

Epple G, Mason JR, Nolte DL, Campbell DL, Effects of predator odors on feeding in the mountain beaver, J Mammalogy, 1993, **74**, 715

Farine JP, Le Quere JL, Duffy J, Semon E, Brossut R, 5-Methyl-4-hydroxy-3(2H)-furanone and 2,5-Dimethyl-4-hydroxy-3(2H)-furanone, two components of the male sex pheromone of *Eurycotis floridana* (Walker) (Insecta, Blattidae, Polyzosteriinae), Biosci Biotechnol Biochem, 1993, **57**, 2026

Fiddaman PJ, Rossall S, The production of antifungal volatiles by *Bacillus subtilis*, J Appl Bacteriol, 1993, **74**, 119

Food Marketing Institute 1993, Supermarket facts, FMI Wash DC

French RC, Volatile chemical germination stimulators of rust and other fungal spores, Mycologia, 1992, **84**, 277

Garcia-Brull PD, Nunez J, Nunez A, The effect of scents on the territorial and aggressive behaviour of laboratory rats, Behavioural Proc, 1993, **29**, 25

Green, AB, Hall MJR, Fergiani M, Chirico J, Husni M, Attracting adult new world screwworm, *Cochliomyia hominivorax*, to odour-baited targets in the field, Med Vet Entomol, 1993, **7**, 59

Hamilton-Kemp TR, McCracken CT, Loughrin JH, Andersen RA, Hildebrand DF, Effects of some natural volatile compounds on the pathogenic fungi *Alternaria alternata* and *Botrytis cinerea*, J Chem Ecol, 1992, **18**, 1083

Hatchwell LC, Overcoming flavor challenges in low-fat frozen dessert, Food Technol, 1994, **48**, 98

Heath RR, Manukian A, Development and evaluation of systems to collect volatile semiochemicals from insects and plants using a charcoal-infused medium for air purification, J Chem Ecol, 1992, **18**, 1209

Herz RS, Cupchik GC, An experimental characterization of odor-evoked memories in humans, Chemical Senses 1992, **17**, 519

Hildebrand DF, Brown GC, Jackson DM, Hamilton-Kemp TR, Effects of some leaf-emitted volatile compunds on aphid population increase, J Chem Ecol, 1993, **19**, 1875

Himejima M, Hobson KR, Otsuka T, Wood DL, Kubo I, Antimicrobial terpenes from oleoresin of ponderosa pine tree *Pinus ponderosa* - A defense mechanism against microbial invasion, J Chem Ecol, 1992, **18**, 1809

Ikawa M, Mosley SP, Barbero LJ, Inhibitory effects of terpene alcohols and aldehydes on growth of green alga *Chlorella pyrenoidosa*, J Chem Ecol, 1992, **18**, 1755

Kang, R, Helms R, Stout MJ, Jaber H, Zhengqing C, Nakatsu T, Antimicrobial activity of the volatile constituents of *Perilla frutescens* and its synergistic effects with polygodial, J Agric Food Chem, 1992, **40**, 2328

Knasko SC, Performance, mood, and health during exposure to intermittent odors, Arch Environm Health, 1993, **48**, 305

Knudsen JT, Tollsten L, Bergstrom LG, Floral scents - A checklist of volatile compounds isolated by headspace techniques, Phytochemistry, 1993, **33**, 253

Koda Y, Kikuta Y, Kitahara T, Nishi T Mori K, Comparisons of various biological activities of stereoisomers of methyl jasmonate, Phytochemistry, 1992, **31**, 1111

Kubo I, Muroi, H, Himejima M, Antibacterial activity against *Streptococcus mutans* of mate tea flavor components, J Agric Food Chem, 1993, **41**, 107

Kumar S, Volatile compunds as short-term preservatives for paddy rice (*Oryza sativa* L) in traditional storage, India, Seed Sci Technol, 1993, **21**, 487

Lanciotti R, Guerzoni ME, Competitive inhibition of *Aspergillus flavus* by volatile metabolites of *Rhizopus arrhizus*, Food Microbiol, 1993, **10**, 367

Lin HC, Phelan PL, Comparison of volatiles from beetle-transmitted *Ceratocystis fagacearum* and 4 noninsect-dependent fungi, J Chem Ecol, 1992, **18**, 1623

Maarse H, Visscher CA (eds) 1992, Volatile Compounds in Food, TNO Zeist, 3rd Supplement

Manley CH, Psychophysiological effect of odor, Crit Revs Food Sci Nutr 1993, **33**, 57

Masson C, Phamdelegue MH, Fonta C, Gascuel J, Arnold G, Nicolas G Kerszberg M, Recent advances in the concept of adaptation to natural odour signals in the honeybee, *Apis mellifera* L, Apidologie, 1993, **24**, 169

Matsuoka H, Dousaki S, Kurata N, Homma T, Nemoto Y, Effect of flavor compunds emitted by white birch on the seed germination of the same plant, J Plant Nutr 1993, **16**, 471

Muroi H, Kubo A, Kubo I, Antimicrobial activity of cashew apple flavor compounds, J Agric Food Chem, 1993, **41**, 1106

Muroi H, Kubo I, Combination effects of antibacterial compounds in green tea flavor against *Streptococcus mutans*, J Agric Food Chem, 1993, **41**, 1102

Nagahara A, Benjamin H, Storkson J, Krewson J, Sheng K, Liu W, Pariza MW, Inhibition of benzo-α-pyrene-induced mouse forestomach neoplas by a principal flavor component of japanese-style fermented soy sauce, Cancer Res, 1992, **52**, 1754

Nagnan P, Cain AH, Rochat D, Extraction and identification of volatile compunds of fermented oil palm sap (palm wine), candidate attractants for the palm weevil, Oleagineux, 1992, **47**, 135

Nowbahari B, Thibout E, Defensive role of allium sulfur compounds for leek moth *Acrolepiopsis assectella* Z (Lepidoptera) against generalist predators, J Chem Ecol, 1992, **18**, 1991

Oh KB, Matsuoka H, Sumita O, Takatori K, Kurata H, Automatic evaluation of antifungal volatile compounds on the basis of the dynamic growth process of a single hypha, Appl Microbiol Biotechnol, 1993, **38**, 790

Olias JM, Sanz LC, Rios JJ, Perez, AG, Inhibitory effect of methyl jasmonate on the volatile ester-forming enzyme system in *Golden Delicious* apples, J Agric Food Chem, 1992, **40**, 266

Oosterhaven K, Hartmans KJ, Huizing HJ, Inhibition of Potato (*Solanum tuberosum*) sprout growth by the monoterpene S-carvone - reduction activity without effect on its messenger RNA level, J Plant Physiol, 1993, **141**, 463

Parliment TH, Croteau R (eds) 1986, Biogeneration of Aromas, ACS Symp Ser 317, ACS Wash DC

Perez RL, GC determination of raspberry ketone and malathione in insect bait concentrates, J Chromatography, 1983, **259**, 176

Phillips TW, Jiang XL, Burkholder WE, Phillips JK, Tran HQ, Behavioral responses to food volatiles by 2 species of stored-product coleoptera, *Sitophilus oryzae* (Curculionidae) and *Tribolium castaneum* (Tenebrionidae), J Chem Ecol, 1993, **19**, 723

Preininger M, Rychlik M, Grosch W, Potent odorants of the neutral fraction of Swiss cheese (*Emmentaler*) In: Maarse H, van der Meij OG (eds) Trends in Flavour Research, Elsevier Amsterdam 1994, 267

Rodriguez-Kabana R, Kloepper JW, Weaver CF, Robertson DG, Control of plant parasitic nematodes with furfural - A naturally occurring fumigant, Nematropica, 1993, **23**, 63

Romel KE, Scott-Dupree CD, Carter MH, Qualitative and quantitative analyses of volatiles and pheromone gland extracts collected from *Galleria mellonella (L)* (Lepidoptera, Pyralidae), J Chem Ecol, 1992, **18**, 1255

Scarpati ML, Loscalzo R, Vita G, *Olea europaea* volatiles attractive and repellent to the olive fruit fly (*Dacus oleae* Gmelin), J Chem Ecol, 1993, **19**, 881

Schellinck HM, Smyth C, Brown R, Wilkinson M, Odor -induced sexual maturation and expression of c-fos in the olfactory system of juvenile female mice, Dev Brain Res, 1993, **74**, 138

Schieberle P, Grosch W, Evaluation of the flavour of wheat and rye bread crush by aroma extract dilution analysis, Z Lebensm Unters Forsch 1987, **185**, 111

Schreier P (ed) 1988, Bioflavour' 87, deGruyter Berlin

Shimoni M, Putievsky E, Ravid U, Reuveni R, Antifugal activity of volatile fractions of essential oils from 4 aromatic wild plants in Israel, J Chem Ecol, 1993, **19**, 1129

Smith RL, Ford RA, Recent progress in the consideration of flavoring ingredients under the food additives amendment, 16. GRAS substances, Food Technol 1993, **47**, 154

Teranishi R, Buttery RG, Sugisawa H (eds) 1993, Bioactive volatile compounds from plants, ACS Symp Ser 525, ACS Wash DC

Thibout E, Guillot JF, Auger J, Microorganisms are involved in the production of volatile kairomones affecting the host seeking behaviour of *Diadromus pulchellus*, a parasitoid of *Acrolepiopsis assectella*, Physiol Entomol, 1993, **18**, 176

Tokuoka K, Mori R, Isshihi K, Inhibitory effects of volatile mustard extract on the growth of yeasts, Nippon Shokuhin Kogyo Gakkaishi, 1992, **39**, 68

Vaughn SF, Gardner HW, Lipoxygenase-derived aldehydes inhibit fungi pathogenic on soybean, J Chem Ecol, 1993, **19**, 2337

Vaughn SF, Spencer GF, Aromatic alcohols and aldehydes as potato tuber sprouting inhibitors, US 723118 A0 911201, 1991

Vaughn SF, Spencer GF, Shasha BS, Volatile compounds from raspberry and strawberry fruit postharvest decay fungi, J Food Sci, 1993, **58**, 793

Vaughn SF, Spencer GF, Volatile monoterpenes as potential parent structures for new herbicides, Weed Sci, 1993, **41**, 114

Ware AB, Kaye PT, Compton SG, Vannoort S, Fig volatiles - their role in attracting pollinators and maintaining pollinator specificity, Plant Syst Evolution, 1993, **186**, 147

Weiler EW, Albrecht T, Groth B, Xia ZQ, Luxem M, Liss H, Andert L, Spengler P, Evidence for the involvement of jasmonates and their octadecanoid precursors in the tendril coiling response of *Bryonia dioica*, Phytochemistry, 1993, **32**, 591

Weurman C (ed)1963, Lists of Volatile Compounds in Food, TNO Zeist

Whitman DW, Eller FJ, Orientation of *Microplitis croceipes* (Hymenoptera, Braconidae) to green leaf volatiles - Dose-response curves, J Chem Ecol, 1992, **18**, 1743

Zeringue HJ, Effects of C6 to C10 alkenals and alkanals on eliciting a defence response in the developing cotton boll, Phytochemistry, 1992, **31**, 2305

Zheng GQ, Kenney PM, Lam LKT, Sesquiterpenes from clove (*Eugenia caryophyllata*) as potential anticarcinogenic agents, J Nat Prod, 1992, **55**, 999

Zhu YC, Keaster AJ, Gerhardt KO, Field observations on attractiveness of selected blooming plants to noctuid moths and electroantennogram responses of black cutworm (Lepidoptera, noctuidae) moths to flower volatiles, Environm Entomol, 1993, **22**, 162

Chapter 2

Ames JM, Elmore JS, Aroma components of yeast extracts, Flavour Fragrance J, 1992, **7**, 89

Avedovech RM, McDaniel MR, Watson BT, Sandine WE, An evaluation of combinations of wine yeast and *Leuconostoc oenos* strains in malolactic fermentation of *Chardonnay* wine, Am J Enol Vitic, 1992, **43**, 253

Babel W, Brinkmann U, Muller RH, The auxiliary substrate concept - an approach for overcoming limits of microbial performances, Acta Biotechnol, 1993, **13**, 211

Babuchowski A, Hammond EG, Glatz BA, Survey of propionibacteria for ability to produce propionic and acetic acids, J Food Protec, 1993, **56**, 493

Bakoyianis V, Kana K, Kalliafas A, Koutinas AA, Low-temperature continuous wine making by kissiris-supported biocatalyst: Volatile byproducts, J Agric Food Chem 1993, **41**, 465

Bassit N, Boquien CY, Picque D, Corrieu G, Effect of initial oxygen concentration on diacetyl and acetoin production by *Lactococcus lactis* subsp. *lactis* biovar *diacetylactis* , Appl Environm Microbiol, 1993, **59**, 1893

Begin A, Beaulieu Y, Goulet J, Castaigne F, Whey fermentation by *Propionibacterium shermanii* immobilized in different gels, Milchwissensch - Milk Sci Intern, 1992, **47**, 411

Beneke ES, Stevenson KE, Classification of food and beverage fungi, In: Food and Beverage Mycology, Beuchat LR (ed) van Nostrand Reinhold New York 1987, 1

Berdagué JL, Monteil P, Montel MC, Talon R, Effects of starter cultures on the formation of flavour compunds in dry sausage, Meat Sci, 1993, **35**, 275

Berger RG, Macku C, German JB, Shibamoto T, Isolation and identification of dry salami volatiles, J Food Sci, 1990, **55**, 1239

Besancon X, Smet C, Chabalier C, Rivemale M, Reverbel JP, Ratomahenina R, Galzy P, Study of surface yeast flora of *Roquefort* cheese, Intern J Food Microbiol, 1992, **17**, 9

Beuchat LR, Traditional fermented food products In: Food and Beverage Mycology, Beuchat LR (ed) van Nostrand Reinhold New York 1987, 269

Blanch, GP, Tabera J, Sanz J, Herraiz M, Reglero G, Volatile composition of vinegars - simultaneous distillation extraction and gas chromatographic mass spectrometric analysis, J Agric Food Chem, 1992, **40**, 1046

Boraam F, Faid M, Larpent JP, Breton A, Lactic acid bacteria and yeast associated with traditional moroccan sour-dough bread fermentation, Sci Aliments, 1993, **13**, 501

Borcakh M, Özay G, Alperden I, Fermentation of Turkish black olives with traditional and aerated systems, In: Food Flavors, Ingredients and Composition, Charalambous G (ed) Elsevier Amsterdam 1993, 265

Brunerie P, Benda I, Bock G, Schreier P, Bioconversion of monoterpene alcohols and citral by *Botrytis cinerea,* In: Bioflavour '87 Schreier P (ed) de Gruyter Berlin 1988, 435

Cachon R, Divies C, Localization of *Lactococcus lactis* ssp. *lactis* biovar *diacetylactis* in alginate gel beads affects biomass density and synthesis of several enzymes involved in lactose and citrate metabolism, Biotechnol Tech, 1993, **7**, 453

Champagne CP, Gaudy C, Poncelet D, Neufeld RJ, *Lactococcus lactis* release from calcium alginate beads, Appl Environm, 1992, **58**, 1429

Chasco J, Beriain MJ, Bello J, A study of changes in the fat content of some varieties of dry sausage during the curing process, Meat Sci, 1993, **34**, 191

Chatonnet P, Dubourdieu D, Boidron JN, Livigne DV, Synthesis of volatile phenols by *Saccharomyces cerevisiae* in wines, J Sci Food Agric, 1993, **62**, 191

Chatonnet P, Dubourdieu D, Boidron JN, Pons M, The origin of ethylphenols in wines, J Sci Food Agric, 1992, **60**, 165

Collar C, Mascaros AF, Debarber CB, Amino acid metabolism by yeasts and lactic acid bacteria during bread dough fermentation, J Food Sci, 1992, **57**, 1423

Collin S, Osman M, Delcambre S, Elzayat AI, Dufour JP, Investigation of volatile flavor compounds in fresh and ripened domiati cheeses, J Agric Food Chem, 1993, **41**, 1659

Croizet F, Denoyer C, Tran N, Berdagué, Les composés volatils du saucisson sec évolution au cours de la maturation, Viandes Prod. Carnés, 1992, **13**, 167

Delteil D, Jarry JM, Characteristic effects of two strains of enological yeasts on the composition of volatile compounds in *Chardonnay* wines, Rev Fr Oenol, 1991, **132**, 41

Demacias MEN, Romero NC, Apella MC, Gonzales SN, Oliver G, Prevention of infections produced by *Escherichia coli* and *Listeria monocytogenes* by feeding milk fermented with *Lactobacilli,* J Food Protec, 1993, **56**, 401

Demuyakor B, Ohta Y, Characteristics of single and mixed culture fermentation of pito beer, J Sci Food Agric, 1993, **62**, 401

Deshpande MV, Ethanol production from cellulose by coupled saccharification fermentation using *Saccharomyces cerevisiae* and cellulase complex from *Sclerotium rolfsii uv*-8 mutant, Appl Biochem Biotechnol, 1992, **36**, 227

Dick KJ, Molan PC, Eschenbruch R, The isolation from *Saccharomyces cerevisiae* of 2 antibacterial cationic proteins that inhibit malolactic bacteria, Vitis, 1992, **31**, 105

Drawert F, The chemistry of winemaking as a biological-technological sequence, In: Chemistry of Winemaking, Webb AD (ed) ACS Symp Ser 137, ACS Wash DC 1974, 1

Dupuis C, Boyaval P, Esterase activity of dairy *Propionibacterium,* Lait, 1993, **73**, 345

Ezeogu LI, Emeruwa AC, High level ethanol-tolerant *Saccharomyces* from Nigerian palm wine, Biotechnol Letters, 1993, **15**, 83

Ezzat N, Elsoda M, Elshafei H, Olson NF, Cell-wall associated peptide hydrolase and esterase activities in several cheese-related bacteria, Food Chemistry, 1993, **48**, 19

Fernandez-Garcia E, Olano A, Cabezudo D, Martin-Alvarez PJ, Ramos M, Accelerated ripening of manchego type cheese by added commercial enzyme preparation from *Aspergillus oryzae,* Enz Microb Technol, 1993a, **15**, 519

Fernandez-Garcia E, Reuter H, Prokopek D, Olano A, Ramos M, Effect of enzyme additi-on on the manufacture of spanish hard cheese from milk concentrated by ultrafiltration ripening of cheeses, Kieler Milchwirtsch Forschungsber, 1993b, **45**, 301

Fischer U, PhD Thesis Universität Hannover 1994

Fitzgerald RJ, Doonan S, McKay LL, Cogan TM, Intracellular pH and the role of D-lactate dehydrogenase in the production of metabolic and products by *Leuconostoc lactis* , J Dairy Res, 1992, **59**, 359

Fourcassie P, Makagakabindamassard E, Belarbi A, Maujean A, Growth, D-glucose uti-lization and malolactic fermentation by *Leuconostoc oenos* strains in 18 media deficient in one amino acid, J Appl Bacteriol, 1992, **73**, 489

Frasse P, Lambert S, Richard-Molard D, Chiron H, The influence of fermentation on volatile compounds in French bread dough, Food Sci Technol, 1993, **26:2**, 126

Frezier V, Dubourdieu D, Ecology of yeast strain *Saccharomyces cerevisiae* during spon-taneous fermentation in a bordeaux winery, Am J Enol Viticult, 1992, **43**, 375

Fukaya M, Park YS, Toda K, Improvement of acetic acid fermentation by molecular breeding and process development - review, J Appl Bacteriol, 1992, **73**, 447

Fukuda T, Sanmoto H, Hiramatsu M, The production of alcoholic beverages with high flavor from saccharified solutions 3. The influence of stirring on the formation of flavor components during fermentation of enzyme-saccharified solution, Nippon Jozo Kyokaishi, 1991, **86**, 684

Garcia ML, Selgas MD, Fernandez M, Ordonez JA, Microorganisms and lipolysis in the ripening of dry fermented sausages, Intern J Food Sci Technol, 1992, **27**, 675

Garriga M, Hugas M, Aymerich T, Monfort JM, Bacteriocinogenic activity of *Lactobacilli* from fermented sausages, J Appl Bacteriol, 1993, **75**, 142

Geisen R, Fungal starter cultures for fermented foods - molecular aspects, Food Sci Tech-nol, 1993, **4**, 251

Giudici P, Zambonelli C, Kunkee RE, Increased production of n-propanol in wine by yeast strains having an impaired ability to form hydrogen sulfide, Am J Enol Vitic, 1993, **44**, 17

Goncalves LMD, Barreto MTO, Xavier AMBR, Carrondo MJT, Klein J, Inert Supports for lactic acid fermentation - a technological assessment, Appl Microbiol Biotechnol, 1992, **38**, 305

Gonzalez JF, Fernandez AG, Garcia PG, Balbuena MB, Quintana MCD, Characteristics of the fermentation process that occurs during the storage in brine of hojiblanca cultivar, used to elaborate ripe olives, Grasas Y Aceites, 1992, **43**, 212

Grando MS, Versini G, Nicolini G, Mattivi F, Selective use of wine yeast strains having different volatile phenols production, Vitis, 1993, **32**, 43

Groboillot AF, Champagne CP, Darling GD, Poncelet D, Neufeld RJ, Membrane forma-tion by interfacial cross-linking of chitosan for microencapsulation of *Lactococcus lactis*, Biotechnol Bioengin, 1993, **42**, 1157

Grosch W, Schieberle P, Bread In: Volatile Compound in Foods and Beverages, Maarse H (ed) Dekker New York 1991, 41

Guichard E, Etievant P, Henry R, Mosandl A, Enantiomeric ratios of pantolactone, solero-ne, 4-carboethoxy-4-hydroxy-butyrolactone and of sotolon, a flavour impact compound of flor-sherry and botrytized wines, Z Lebensm Unters Forsch 1992, **195**, 540

Halm M, Lillie A, Soerensen AK, Jakobsen M, Microbiological and aromatic characteristics of fermented maize doughs for kenkey production in Ghana, Int J Food Microbiol, 1993, **19**, 135

Hamad SH, Bocker G, Vogel RF, Hammes WP, Microbiological and chemical analysis of fermented sorghum dough for kisra production, Appl Microbiol Biotechnol, 1992, **37**, 728

Hammes WP, Bacterial starter cultures in food production, Food Biotechnol, 1990, **4**, 383

Hammes WP, Fermentation of non-dairy foods, Food Biotechnol, 1991, **5**, 293

Hammes WP, Tichaczek PS, The potential of lactic acid bacteria for the production of safe and wholesome food, Z Lebensm Unters Forsch, 1994, **198**, 193

Hammond EG, The flavor of dairy products In: Flavor Chemistry of Lipid Foods, Min DB, Smouse TH (eds) AOCS 1989, 222

Hansen, B, Hansen Å, Volatile compounds in wheat sourdoughs produced by lactic acid bacteria and sourdough yeasts, Z Lebensm Unters Forsch, 1994, **198**, 202

Heidlas J, Tressl R, Purification and characterization of a (R)-2,3-butanediol dehydrogenase from *Saccharomyces cerevisiae*, Arch Microbiol, 1990, **154**, 267

Hock R, Benda I, Schreier P, Formation of terpenes by yeasts, Z Lebensm Unters Forsch, 1984, **179**, 450

Holloway P, Subden RE, Volatile metabolites produced in a *Riesling* must by wild yeast isolates, Can Inst Food Sci Technol J, 1991, **24**, 57

Holm CS, Aston JW, Doglas K, The effects of the organic acids in cocoa on the flavour of chocolate, J Sci Food Agric, 1993, **61**, 65

Hugenholtz J, Perdon L, Abee T, Growth and energy generation by *Lactococcus lactis* subsp *lactis* biovar *diacetylactis* during citrate metabolism, Appl Environm Microbiol, 1993, **59**, 4216

Hupf H, Schmid W, Wein: Über die Stereoisomeren des 2,3-Butandiols, Dtsch Lebensm Rdsch,1994, **1**, 1

Hwang GR, Chou CC, Production of some flavor components by *Streptococcus faecium* and *Torulaspora delbrückii* in koji-extract medium and tou-pan-chiang, J Chin Agric Chem Soc, 1991, **29**, 475

Hwang HJ, Vogel RF, Hammes WP, Development of mould cultures for sausage fermentation - characterisation and toxicological assessment, Fleischwirtsch, 1993, **73**, 89 and 327

Hyndman CL, Groboillot AF, Poncelet D, Champagne CP, Neufeld RJ, Microencapsulation of *Lactococcus lactis* within cross-linked gelatin membranes, J Chem Technol Biotechnol, 1993, **56**, 259

Iida T, Sakamoto M, Izumida H, Akagi Y, Characteristics of *Zymomonas mobilis* immobilized by photocrosslinkable resin in ethanol fermentation, J Ferment Bioengin, 1993, **75**, 28

Imhof R, Bosset JO, Relationships between micro-organisms and formation of aroma compounds in fermented dairy products, Z Lebensm Unters Forsch, 1994, **198**, 267

Ito, H, Toeda K, Preparation of salt seasoning by fermentation of molasses, JP 05056764 A2 930309, 1993, Iwasaki K, Nakajima M, Sasahara H, Rapid continuous lactic acid fermentation by immobilised lactic acid bacteria for soy sauce production, Proc Biochem, 1993, **28**, 39

Iwasaki KI, Nakajima M, Sasahara H, Porous alumina beads for immobilization of lactic acid bacteria and its application for repeated-batch fermentation in soy sauce production, J Ferment Bioengin, 1992, **73**, 375

Jackson TC, Acuff GR, Sharp TR, Savell JW, Volatile compounds on sterile pork loin tissue inoculated with *Lactobacillus plantarum* and *Lactobacillus fermentum*, J Food Sci, 1992, **57**, 783

Javanainen P, Linko YY, Factors affecting rye sour dough fermentation with mixed-culture pre-ferment of lactic and propionic acid bacteria, J Cereal Sci, 1993, **18**, 171

Jeppesen VF, Huss HH, Antagonistic activity of 2 strains of lactic acid bacteria against *Listeria monocytogenes* and *Yersinia enterocolitica* in a model fish product at 5-degrees-C, Intern J Food Microbiol, 1993, **19**, 179

Junker M, Porobic R, Sieber W, Linhard O, Knauf HJ, Rohwurstherstellung. Beschreibung einer neuen Starterkultur mit *Pediococcus pentosaceus,* Fleischwirtsch, 1993, **73**, 325

Kaminarides SE, Anifantakis EM, Balis C, Changes in *Kopanisti* cheese during ripening using selected pure microbial cultures, J Sci Dairy Technol, 1992, **45**, 56

Kaneda H, Kano Y, Sekine T, Ishii S, Takahashi K, Koshino S, Effect of pitching yeast and wort preparation on flavor stability of beer, J Ferment Bioengin, 1992, **73**, 456

Kanematsu Y, Kasahara M, Hiraguri Y, Honkawa Y, Production of a shiro soy sauce like seasoning by bioreactors, Nippon Shoyu Kenkyusho Zasshi, 1992, **18**, 260

Kelly WJ, Huang CM, Asmundson RV, Comparison of *Leuconostoc oenos* strains by pulsed-field gel electrophoresis, Appl Environm Microbiol, 1993, **59**, 3969

Kida K, Nishimura K, Nakagawa M, So Y, Production of shochu from crushed rice by noncooking fermentation with saccharifying enzymes, Nippon Jozo Kyokaishi, 1991, **86**, 962

Kim WJ, Bacteriocins of lactic acid bacteria - their potentials as food biopreservative, Food Rev Intern, 1993, **9**, 299

Kirk LA, Doelle HW, Rapid ethanol production from sucrose without by-product formation, Biotechnol Letters, 1993, **15**, 985

Kishimoto M, Shinohara T, Soma E, Goto S, Selection and fermentation properties of cryophilic wine yeasts , J Ferment Bioengin, 1993, **75**, 451

Klaver FAM, Kingma F, Timmer JMK, Weerkamp AH, Interactive fermentation of milk by means of a membrane dialysis fermenter - buttermilk, Netherl Milk Dairy J, 1992, **46**, 19 and 31

Kneifel W, Ulberth F, Erhard F, Jaros D, Aroma profiles and sensory properties of yogurt and yogurt-related products 1. screening of commercially available starter cultures, Milchwissensch - Milk Sci Intern, 1992, **47**, 362

Kruger L, Pickerell ATW, Axcell B, The sensitivity of different brewing yeast strains to carbon dioxide inhibition: Fermentation and production of flavor-active volatile compunds, J Inst Brewing, 1992, **98**, 133

Kunze G, Kunze I, Barner A, Schulz R, Genetical and biochemical characterization of *Saccharomyces cerevisiae* industrial strains, Fresenius J Anal Chem, 1993, **346**, 868

Kuriyama H, Mahakarnchanakul W, Matsui, S, Kobayashi H, The effects of pCO_2 on yeast growth and metabolism under continuous fermentation, Biotechnol Letters, 1993, **15**, 189

Kuwabara H, Oguri I, Baba S, Manufacture of alcoholic beverages with fungi, JP 05056774 A2 930309, 1993

Laplace JM, Delgenes JP, Moletta R, Navarro JM, Effects of culture conditions on the co-fermentation of a glucose and xylose mixture to ethanol by a mutant of *Saccharomyces diastaticus* associated with *Pichia stipitis*, Appl Microbiol Biotechnol, 1993, **39**, 760

Law J, Fitzgerald GF, Daly C, Fox PF, Farkye NY, Proteolysis and flavor development in cheddar cheese made with the single starter strains *Lactococcus lactis* ssp *lactis* UC317 or *Lactococcus lactis* ssp *cremoris* HP, J Dairy Sci, 1992, **75**, 1173

Laye I, Karleskind D, Morr CV, Chemical, microbiological and sensory properties of plain nonfat yogurt, J Food Sci, 1993, **58**, 991

Lazos ES, Aggelousis G, Bratakos M, The fermentation of trahanas - a milk-wheat flour combination, Plant Foods Human Nutr, 1993, **44**, 45

le Roux M, van Vuuren HJJ, Dicks LMT, Loos MA, Simple headspace concentration trap for capillary gas chromatographic analysis of volatile metabolites of *Leuconostoc oenos*, System Appl Microbiol, 1989, **11**, 176

Lee SK, Johnson ME, Marth EH, Characteristics of reduced-fat cheddar cheese made with added *Micrococcus species* L13, Food Sci Technol, 1992a, **25**, 552

Lee, CH, Min KC, Souane M, Chung MJ, Mathiasen TE, Adlernisse J, Fermentation of prefermented and extruded rice flour by the lactic acid bacteria from sikhae, Food Biotechnol, 1992b, **6**, 239

Lee, SW, Yajima M, Tanaka H, Use of food additives to prevent contamination during fermentation using a co-immobilized mixed culture system, J Ferment Bioengin, 1993, **75**, 389

Leroi F, Pidoux M, Detection of interactions between yeasts and lactic acid bacteria isolated from sugary kefir grains, J Appl Bacteriol, 1993, **74**, 48 and 54

Lewis VP, Yang ST, Propionic acid fermentation by *Propionibacterium acidipropionici* - effect of growth substrate, Appl Microbiol Biotechnol, 1992, **37**, 437

Li Q, Studies on the flavor compounds of soymilk yogurt from lactic acid bacteria fermentation, Shipin Yu Fajiao Gongye, 1986, **2**, 1

Linden T, Peetre J, Hahn-Hagerdal B, Isolation and characterization of acetic acid-tolerant galactose-fermenting strains of *Saccharomyces cerevisiae* from a spent sulfite liquor fermentation plant, Appl Environm Microbiol, 1992, **58**, 1661

Longo E, Velazquez JB, Sieiro C, Cansado J, Calo P, Villa TG, Production of higher alcohols, ethyl acetate, acetaldehyde and other compunds by 14 *Saccharomyces cerevisiae* wine strains isolated from the same region (Salnes, NW spain) World J Microbiol Biotechnol, 1992, **8**, 539

Lues JFR, Viljoen BC, Miller M, Prior BA, Interaction of non-culture microbial flora on dough fermentation, Food Microbiol, 1993, **10**, 205

Makanjuola DB, Tymon A, Springham DG, Some effects of lactic acid bacteria on laboratory-scale yeast fermentations, Enzyme Microb Technol, 1992, **14**, 350

Marchesini B, Bruttin A, Romailler N, Moreton RS, Stucchi C, Sozzi T, Microbiological events during commercial meat fermentations, J Appl Bacteriol, 1992, **73**, 203

Marshall VM, Lactic acid bacteria: starter for flavour, FEMS Microbiol Rev, 1987, **46**, 327

Marshall VM, Starter cultures for milk fermentation and their characteristics, J Sci Dairy Technol, 1993, **46**, 49

Martens H, Dawoud E, Verachtert H, Synthesis of aroma compunds by wort *Enterobacteria* during the 1st stage of lambic fermentation, J Inst Brewing, 1992, **98**, 421

Martinez-Force E, Benitez T, Changes in yeast amino acid pool with respiratory versus fermentative metabolism, Biotechnol Bioengin, 1992, **40**, 643

Masschelein CA, Recent and future developments of fermentation technology and fermenter design in brewing In: Biotechnology Applications in Beverage Production, Cantarelli C, Lanzanni G (eds) Elsevier London 1989, 77

Mateo JJ, Jimenez M, Huerta T, Pastor A, Comparison of volatile produced by four *Saccharomyces cerevisiae* strains isolated from *Monastrell* musts, Am J Enol Vitic, 1992, **43**, 206

Mateo JJ, Jimenez M, Huerta T, Pastor A, Contribution of different yeast isolated from musts of *Monastrell* grapes to the aroma of wine, Int J Food Microbiol, 1991, **24**, 153

Matsuura K, Hirotsune M, Hamachi M, Nunokawa Y, Thermal control strategy for isoamyl acetate formation in sake brewed with a saccharified rice solution, J Ferment Bioengin, 1992, **74**, 112

Mauricio JC, Moreno JJ, Valero EM, Zea L, Medina M, Ortega JM, Ester formation and specific activities of *vitro* alcohol acetyltransferase and esterase by *Saccharomyces cerevisiae* during grape must fermentation, J Agric Food Chem, 1993, **41**, 2086

Mauricio JC, Salmon JM, Apparent loss of sugar transport activity in *Saccharomyces cerevisiae* max mainly account for maximum ethanol production during alcoholic fermentation, Biotechnol Letters, 1992, **14**, 577

McFeeters RF, Single-injection HPLC analysis of acids, sugars, and alcohols in cucumber fermentations, J Agric Food Chem, 1993, **41**, 1439

Meraz M, Shirai K, Larralde P, Revah S, Studies on the bacterial acidification process of cassava (*Manihot esculenta*), J Sci Food Agric, 1992, 60, 457

Meurer P, Gierschner K, Occurrence and effect of indigenous and eventual microbial enzymes in lactic acid fermented vegetables, Acta Alimentaria, 1992, **21**, 171

Michel RH, McGovern PE, Badler VR, Chemical evidence for ancient beer, Nature , 1992, **360**, 24

Minarik E, Jungova O, Effect of yeast ghost and cellulose preparations on different yeast species occurring in must and wine, Wein-Wiss, 1992, **47**, 140

Montano A, Sanchez AH, Decastro A, Controlled Fermentation of Spanish-type green olives, J Food Sci, 1993, 58, 842

Muir DD, Banks JM, Hunter EA, Sensory changes during maturation of fat-reduced cheddar cheese - effect of addition of enzymically active attenuated starter cultures, Milk Sci Intern, 1992, **47**, 218

Muramatsu S, Ito N, Sano Y, Uzuka Y, Soy sauce manufacture from koji autolyzed at high temperature. Intermediate-scale fermentation test, Nippon Jozo Kyokaishi, 1992, **87**, 538

Murti TW, Bouillanne C, Landon M, Desmazeaud MJ, Bacterial growth and volatile compunds in yoghurt-type products from soymilk containing *Bifidobacterium* ssp, J Food Sci, 1993a, **58**, 153

Murti TW, Lamberet G, Bouillanne C, Desmazeaud MJ, Landon M, Lactobacilli growth in soy milk. Effects on viscosity, volatile compounds and proteolysis, Sci Aliments, 1993b, **13**, 491

Murti TW, Roger S, Bouillanne C, Landon M, Desmazeaud M, Growth of *Bifidobacterium* sp-CNRZ 1494 in soy-extract and cow milk effects on aroma compounds, Sci Aliments, 1992, **12**, 429

Nagodawithana T, Yeast-derived flavors and flavor enhancers and their probable mode of action, Food Technol, 1992, 138

Nsofor LM, Nsofor ON, Nwachukwu KE, Soya-yoghurt starter culture development from fermented tropical vegetables, J Sci Food Agric, 1992, **60**, 515

O'Reilly A, Scott JA, Use of an ion-exchange sponge to immobilise yeast in high gravity apple based (cider) alcoholic fermentations, Biotechnol Letters, 1993, **15**, 1061

Ohta K, Hamada S, Nakamura T, Production of high concentrations of ethanol from inulin by simultaneous saccharification and fermentation using *Aspergillus niger* and *Saccharomyces cerevisiae*, Appl Environm Microbiol, 1993, **59**, 729

Okonogi S, Tomita M, Shimamura S, Toyama K, Myagawa H, Fujimoto M, Fermentation flavors manufacture, JP 04169166 A2 920617, 1992

Olson NF, The impact of lactic acid bacteria on cheese flavor, FEMS Microbiol Rev, 1990, **87**, 131

Parascandola P, Dealteriis E, Farris GA, Budroni M, Scardi V, Behaviour of grape must ferment *Saccharomyces cerevisiae* immobilized within insolubilized gelatin, J Ferment Bioengin, 1992, **74**, 123

Pardo I, Zuniga M, Lactic acid bacteria in spanish red rose and white musts and wines under cellar conditions, J Food Sci, 1992, **57**, 392

Park SK, PhD Thesis University of California 1993

Perez SR, Miura H, Mikami M, Sekikawa M, Action of isolated *Micrococcus* sp, *Pediococcus* sp and *Lactobacillus* sp fermented dry sausage, Obihiro Chikusan Daigaku Gakujutsu Kenkyu Hokoku, Dai-1-Bu, 1992, **17**, 367

Pfleger R, Results and consequences of the culture programm for hard cheese, Milchw Ber Bundesanst Wolfpassing Rotholz, 1992, **110**, 11

Preininger M, Rychlik M, Grosch W, Potent odorants of the neutral volatile fraction of Swiss cheese (*Emmentaler*) In: Trends in Flavour Research, Maarse H van der Heij GD (eds) Elsevier Amsterdam 1994, 267

Prevost H, Divies C, Cream fermentation by a mixed culture of Lactococci entrapped in 2-layer calcium alginate gel beads, Biotechnol Let, 1992, **14**, 583

Ranadive KS, Pai JS, Flavor production by yeasts: Isolation and screening of *H. anomala* and *S. cerevisiae*, PAFAI J, 1991, **13**, 31

Reiss J, Miso from peas (*Pisum sativum*) and beans (*Phaseolus vulgaris*) of domestic origin fermented foods from agricultural products in europe 2, Z Ernährungswissensch, 1993a, **32**, 237

Reiss J, Preparation of tempeh from domestic peas, Dtsch Lebensm Rundsch 1993b, **89**, 147

Renger RS, Vanhateren SH, Luyben KCAM, The formation of esters and higher alcohols during brewery fermentation - The effect of carbon dioxide pressure , J Inst Brewing, 1992, **98**, 509

Requena T, Pelaez C, Fox PF, Peptidase and Proteinase activity of *Lactococcus lactis*, *Lactobacillus casei* and *Lactobacillus plantarum*, Z Lebensm Unters Forsch, 1993, **196**, 351

Ribereau-Gayon P, Effect of yeast strains on wine flavor, C R Acad Agric Fr, 1993, **79**, 73

Richter K, Ruhlemann I, Berger R, High-performance fermentation with lactic acid bacteria entrapped in pectate gel - immobilizates with enhanced lactate formation activity, Acta Biotechnol, 1992, **12**, 229

Romano P, Suzzi G, Comi G, Zironi R, Higher alcohol and acetic acid production by apiculate wine yeasts, J Appl Bacteriol, 1992, **73**, 126

Rosi I, Bertuccioli M, Influences of lipid addition on fatty acid composition of *Saccharomyces cerevisiae* and aroma characteristics of experimental wines, J Inst Brewing, 1992, **98**, 305

Rosi I, Contini M, Bertuccioli M, Relationship between enzymatic activities of wine yeasts and aroma compound formation In: Flavors and Off Flavors, Charalambous G (ed) Elsevier Amsterdam 1990, 24

Roudotalgaron F, Lebars D, Einhorn J, Adda J, Gripon JC, Flavor constituents of aqueous fraction extracted from comte cheese by liquid carbon dioxide, J Food Sci, 1993, **58**, 1005

Russell, I, Graham G St, Contribution of yeast and immobilization technology to flavor development in fermented beverages, Food Technology, 1992, 146

Saigusa T, Harada M, Okamura S, Shinohara T, Study on the control of shochu flavour 2. factors affecting the formation of isoamyl acetate during all-koji shochu fermentation, Seibutsu-Kogaku Kaishi - J Soc Ferment Bioengin, 1993b, **71**, 383

Saigusa T, Harada M, Shinohara T, Control of flavor formation of changing the time of koji culture during all-koji shochu fermentation, Seibutsu Kogaku Kaishi, 1993a, **71**, 105

Sakaguchi M, Hirose T, Nakatani K, Onishi M, Kumada J, The effect of reduced pressure on the growth of yeast cells and on the production of volatile compunds, Hakko Kogaku Kaishi, 1990, **68**, 261

Sakamoto K, Shimoda M, Osajima Y, Concentration in Porapak Q column of volatile compounds in sake for analysis, Nippon Nogeikagaku Kaishi - J Jap Soci Biosci Biotechnol Agrochem, 1993, **67**, 685

Samah OA, Ibrahim N, Alimon H, Karim MIA, Fermentation studies of stored cocoa beans, World J Microbiol Biotechnol, 1993, **9**, 603

Sanceda NG, Kurata T, Suzuki Y, Arakawa N, Oxygen effect on volatile acids formation during fermentation in manufacture of fish sauce, J Food Sci, 1992, **57**, 1120

Sanni AI, The need for process optimization of African fermented foods and beverages - Review, Intern J Food Microbiol, 1993, **18**, 85

Sasaki M, Mori S, The flavor of shoyu, Nippon Jozo Kyokaishi, 1991, **86**, 913

Schieberle P, Formation of furaneol in heat-processed foods In: Flavor Precursors, Teranishi R, Takeoka GR, Güntert M (eds) ACS Symp Ser 490, ACS Wash 1992, 164

Schieberle, P, Grosch W, Potent odorants of rye bread crust - differences from the crumb and from wheat bread crust, Z Lebensm Unters Forsch, 1994, **198**, 282

Selgas D, Garcia L, Defernando GG, Ordonez JA, Lipolytic and proteolytic activity of Micrococci isolated from dry fermented sausages, Fleischwirtsch, 1993, **73**, 1164

Shindo S, Murakami J, Koshino S, Control of acetate ester formation during alcohol fermentation with immobilized yeast, J Ferment Bioengin, 1992, **73**, 370

Shindo S, Sahara H, Koshino S, Tanaka H, Control of diacetyl precursor [alpha-acetolactate] formation during alcohol fermentation with yeast cells immobilized in alginate fibers with double gel layers, J Ferment Bioengin, 1993, **76**, 199

Slaughter, JC, Nomura T, Autocatalytic degradation of proteins in extracts of a brewing strain of *Saccharomyces cerevisiae* - the role of endoproteinases and exopeptidases, Appl Microbiol Biotechnol, 1992, **37**, 638

Sousa MJ, Teixeira JA, Mota M, Must deacidification with an induced flocculant yeast strain of *Schizosaccharomyces pombe*, Appl Microbiol Biotechnol, 1993, **39**, 189

Stashenko H, Macku C, Shibamoto T, Monitoring volatile chemicals formed from must during yeast fermentation, J Agric Food Chem, 1992, **40**, 2257

Strohmar W, Diekmann H, The microflora of a sourdough developed during extended souring phases, Z Lebensm Unters Forsch, 1992, **194**, 536

Suarez JA, Agudelo J, Characterization of yeast and lactic acid bacteria species in ropy wines, Z Lebensm Unters Forsch, 1993, **196**, 152

Sugawara E, Saiga S, Kobayashi A, Relationships between aroma components and sensory evaluation of miso, J Jap Soci Food Sci Technol, 1992, **39**, 1098

Takatsuji W, Ikemoto S, Skaguchi H, Minami H, Development of a new type of umeshu using immobilized growing yeast cells, Nippon Jozo Kyokaishi, 1992, **87**, 533

Takezaki M, Matsuura K, Hirotsune M, Hamachi M, Effects of solids on the growth of yeast, Nippon Jozo Kyokaishi, 1993, **88**, 319

Tamada M, Begum AA, Sadi S, Production of L(+)-lactic acid by immobilized cells of *Rhizopus oryzae* with polymer supports prepared by gamma-ray induced polymerization, J Ferment Bioengin, 1992, **74**, 379

Tanaka T, Shoji Z, Analysis of volatile compunds in the natto-fermentating room by gas-chromatography mass-spectrometry, J Jap Soci Food Sci Technol - Nippon Shokuhin Kogyo Gakkaishi, 1993, **40**, 656

Teramoto Y, Okamoto K, Kayashima S, Ueda S, Rice wine brewing with sprouting rice an barley malt, J Ferment Bioengin, 1993, **75**, 460

Thomas CS, Boulton RB, Silacci MW, Gubler WD, The effect of elemental sulfur, yeast strain, and fermentation medium on hydrogen sulfide production during fermentation, Am J Enol Vitic, 1993, **44**, 211

Thomas KC, Ingledew WM, Production of 21-% (v/v) ethanol by fermentation of very high gravity (vgh) wheat mashes, J Ind Microbiol, 1992, **10**, 61

Tomlins KI, Baker DM, Daplyn P, Adomako D, Effect of fermentation and drying practices on the chemical and physical profiles of Ghana cocoa, Food Chem, 1993, **46**, 257

Torner MJ, Martínez-Anaya MA, Antuña B, Benedito de Barber C, Headspace flavour compounds produced by yeasts and lactobacilli during fermentation of preferments and bread doughs, J Food Microbiol, 1992, **15**, 145

Trepanier G, Elabboudi M, Lee BH, Simard RE, Accelerated maturation of cheddar cheese - microbiology of cheeses supplemented with *Lactobacillus casei* subsp *casei* L2A and influence of added lactobacilli and commercial protease on composition and texture, J Food Sci, 1992, **57**, 345 and 898

van Vuuren HJJ, Dicks LMT, *Leuconostoc oenos* - a review, Am J Enol Vitic, 1993, **44**, 99

Verhue WMM, Tjan SB, Verrips CT, Van Schie BJ, Preparation of an aroma product containing alpha-acetolactic acid, EP Appl 91--202042 910809, 1992

Visser S, Proteolytic enzymes and their relation to cheese ripening and flavor: an overview, J Dairy Sci, 1993, **76**, 329

Vogel RF, Lohmann M, Nguyen M, Weller AN, Hammes WP, Molecular characterization of *Lactobacillus curvatus* and *L. sake* isolated from sauerkraut and their application in sausage fermentations, J Appl Bacteriol, 1993, **74**, 295

Voigt J, Ziehl B, Heinrichs H, Proteolytic formation of cocoa flavour precursors In: Progress in Flavour Precursor Studies, Schreier P, Winterhalter P (eds) Allured Carol Stream 1993, 213

Vösgen W, Rohwurst, bewährte und neue Wege zur Produktion, Fleischwirtsch, 1993, **73**, 723

Walker MD, Simpson, WJ, Production of volatile sulphur compounds by ale and lager brewing strains of *Saccharomyces cerevisiae*, Letters Appl Microbiol, 1993, **16**, 40

Werkhoff P, Bretschneider W, Emberger R, Güntert M, Hopp R, Köpsel M, Recent developments in the sulfur flavor chemistry of yeast extracts, Chem Mikrobiol Technol Lebensm, 1991, **13**, 30

Yang TS, Min DB, Dynamic headspace analysis of volatile compounds of cheddar and swiss cheeses during ripening In: Food Flavors, Ingredients and Composition, Charalambous G (ed) Elsevier Amsterdam 1993, 157

Yankah VV, Ohshima T, Koizumi C, Effects of processing and storage on some chemical characteristics and lipid composition of a Ghanaian fermented fish product, J Sci Food Agric, 1993, **63**, 227

Yoneyama T, Toida J, Baba, S, Studies on using of enzyme preparations for making of miso. 2. effects of proteolytic enzyme preparations and glutaminase on the quality of miso, Nagano-ken Shokuhin Kogyo Shikenjo Kenkyu Hokoku, 1992, **20**, 7

Zironi R, Romano P, Suzzi G, Battistutta F, Comi G, Volatile metabolites produced in wine by mixed and sequential cultures of *Hanseniaspora guilliermondii* or *Kloeckera apiculata* and *Saccharomyces cerevisiae*, Biotechnol Letters, 1993, **15**, 235

Chapter 3

Appler WD, Giamporcaro DE, Regulatory aspects of biotechnology produced ingredients for food application In: Biotechnology and Food Ingredients, Goldberg I, Williams R (eds) van Nostrand Reinhold New York 1991, 537

Chastrette M, Rognon C, Sauvegrain P, Amouroux R, On the role of chirality in structure-odor relationships, Chemical Senses, 1992, **17**, 5

Dhavlikar RS, Albroscheit G, Mikrobiologische Umsetzung von Terpenen: Valencen, Dragoco Report, 1973, **20**, 251

Drawert F, Berger RG, Pflanzliche Zellkulturen in der Lebensmitteltechnologie, In: 45. Disk Tagung, FK der Ernährungsindustrie Hannover, 1987, 93

IOFI, International Organization of the Flavor Industry, Circular Letter 91/6 of July 8, 1991, Geneva

Kollmannsberger H, Berger RG, Industrial recovery of strawberry flavour - fractional distillation vs plate condensation, Z Lebensm Unters Forsch, 1994, **198**, 491

Matheis G, Natürliche Aromen und ihre Rohstoffe, Lebensm Biotechnol, 1989, **6**, 121

Noyori R, Asymmetric terpene synthesis In: Recent Developments in Flavor and Fragrance Chemistry, Hopp R, Mori K (eds) VCh Weinheim 1993, 3

Reineccius GA, Utilizing lost flavor in food manufacturing, Cereal Foods World, 1992, **37**, 392

Tombs MP 1990, Biotechnology in the Food Industry, Open University Press, Milton Keynes

Chapter 4

Abbott N, Etievant P, Issanchou S, Langlois D, Critical evaluation of two commonly used techniques for the treatment of data from extract dilution sniffing analysis, J Agric Food Chem, 1993, **41**, 1698

Abraham B, Krings U, Berger RG, Biotechnologische Produktion von Aromastoffen durch Basidiomyceten, GIT Fachz Lab, 1994, **38**, 370

Abraham B, Krings U, Berger RG, Dynamic extraction, an efficient screening procedure for aroma producing Basidiomycetes, Chem Mikrobiol Technol Lebensm, 1993, **15**, 178

ATCC, The American Type Culture Collection, 12301 Parklawn Dr, Rockville, MD 20825

Berger RG, Dettweiler GR, Kollmannsberger H, Drawert F, The peroxisomal dienoyl-CoA reductase pathway in pineapple (*Ananas comosus*) fruit, Phytochemistry, 1990, **29**, 2069

Braunsdorf R, Hener U, Lehmann D, Mosandl A, Analytical differentiation of natural, fermented, and synthetic (nature-identical) aromas, Part 1, Origin-specific analysis of (E)-alpha (beta)-ionones, Dtsch Lebensm Rdsch, 1991, **87**, 277

Braunsdorf R, Hener U, Przibilla G, Piecha S, Mosandl A, Analytische und technologische Einflüsse auf das $^{13}C/^{12}C$-Isotopenverhältnis von Orangenöl-Komponenten, Z Lebensm Unters Forsch, 1993, **197**, 24

CBS, Centraalbureau voor Schimmelcultures, The Netherlands, POB 273, 3740 AG Baarn

Culp RA, Noakes JE, Determination of synthetic components in flavors by deuterium hydrogen isotopic ratios, J Agric Food Chem, 1992, **40**, 1892

Fronza G, Fuganti C, Grasselli P, Natural abundance ^2H nuclear magnetic resonance study of the origin of (R)-δ-decanolide, J Agric Food Chem, 1993, **41**, 235

Full G, Winterhalter P, Schreier P, Coupled-online high-resolution gas chromatography-FTIR analysis of acetylenic aroma precursors, Labor Praxis, 1991, **15**, 752

Hädrich-Meyer S, Berger RG, Localization of lipolytic and esterolytic activities of *Ty-romyces sambuceus*, a 4-decanolide producing basidiomycete, Appl Microbiol Biotechnol, 1994, 41, 210

Hamberg M, A method for determination of the absolute stereochemistry of alpha,beta-epoxy alcohols derived from fatty acid hydroperoxides, Lipids, 1992, 27, 1042

Hener U, Mosandl A, Zur Herkunftsbeurteilung von Aromen, Dtsch Lebensm Rdsch, 1993, 89, 307

Kapfer GF, Berger RG, Drawert F, Improved production of 4-decanolide by biomass recirculation, Chem Mikrobiol Technol Lebensm, 1991, 13, 1

Kreis P, Mosandl A, Chiral compounds of essential oils. Part XII. Authenticity control of rose oils using enantioselective multidimensional gas chromatography, Flavour Fragrance J, 1992, 7, 199

Lehmann D, Dietrich A, Schmidt S, Dietrich H, Mosandl A, Stereodifferenzierung von $\gamma(\delta)$-Lactonen und (E)-α-Ionon verschiedener Früchte und ihrer Verarbeitungsprodukte, Z Lebensm Unters Forsch, 1993, 196, 207

Mosandl A, Capillary gas chromatography in quality assessment of flavours and fragran-ces, J Chromatography, 1992, 624, 267

Nago H, Matsumoto M, Enantiomeric composition of 5-alkanolides contained in coconut and dairy products, Biosci Biotechnol Biochem, 1993, 57, 427

Reil G, Leupold G, Nitz S, Eluentensteuerung in der Radio-Kapillargaschromatographie mittels eines 'Live-T-Stücks', Chem Mikrobiol Technol Lebensm, 1992, 14, 162

Schmidt G, Full G, Winterhalter P, Schreier P, Synthesis and enantiodifferentiation of isomeric theaspiranes, J Agric Food Chem, 1992, 40, 1188

Schmidt UL, Butzenlechner M, Rossmann A, Schwarz S, Kexel H, Kempe K, Z Lebensm Unters Forsch, 1993, 196, 105

Tateo F, Russo G, Panza L, NMR differentiation between natural and synthetic flavouring substances: method applied to the ethyl butyrate molecule, In: Food Flavors, Ingredients and Composition, Charalambous G (ed) Elsevier Sci Publ 1993, 803

Tiefel P, Berger RG, Volatiles in precursor fed cultures of *Basidiomycetes*, Progress in Flavor Precursor Studies, Hrsg. P.Schreier, P.Winterhalter, Allured Publ., Carol Stream IL, USA 1993, 439

Vorderwülbecke T, Epple M, Cammenga HK, Petersen S, Kieslich K, Racemic lactones from butterfat - An advanced approach that includes stereodifferentiation, JAOCS, 1992, 69, 797

Werkhoff P, Brennecke S, Bretschneider W, Enantiomeric composition of γ-decalactone in raspberry flavor, Chem Mikrobiol Technol Lebensm, 1993a, 15, 47

Werkhoff P, Brennecke S, Bretschneider W, Güntert M, Hopp R, Surburg H, Chirospeci-fic analysis in essential oil, fragrance and flavor research, Z Lebensm Unters Forsch, 1993b, 196, 307

Chapter 5

Abraham B, Berger RG, Higher fungi for generating aroma compounds through novel biotechnologies, J Agric Food Chem, 1994, 42, 2344

Abraham B, Krings U, Berger RG, Biotechnologische Produktion von Aromastoffen durch Basidiomyceten, GIT Fachz Lab, 1994, **38**, 370

Abraham B, Krings U, Berger RG, Dynamic extraction, an efficient screening procedure for aroma producing basidiomycetes, Chem Mikrobiol Technol Lebensm, 1993, **15**, 178

Abraham WR, Arfmann, HA, Stumpf B, Washausen P, Kieslich K, Microbial transformation of some terpenoids and natural compounds In: Bioflavour '87, Schreier P (ed) deGruyter Berlin 1988, 399

Abraham WR, Arfmann HA, Fusalanipyrone, a monoterpenoid from *Fusarium solani*, Phytochemistry, 1988, 10, 3310

Ajinomoto 1993, JP 05227896 after CA 120: 29826 f

Akita Jujo Kasei KK, JP 04045793 of 1992, CA 117: 46742 f

Armstrong DW, Brown LA, Porter S, Rutten R, Biotechnological derivation of aromatic flavour compounds and precursors In: Progress in Flavor Precursor Studies, Schreier P, Winterhalter P (eds) Allured Carol Stream IL 1993, 425

Arnone A, Cardillo R, Nasini G, de Pava OV, Two cinnamic allenic ethers from the fungus *Clitocybe eucalyptorum*, Phytochemistry, 1993, **32**, 1279

Arnone A, Colombo A, Nasini G, Meille SV, Secondary mould metabolites. Eleganthol, a sesquiterpene from *Clitocybe elegans*, Phytochemistry, 1993, **32**, 1493

Badock EC, Preliminary account of the odour of wood-destroying fungi in culture, Trans British Mycol Sci, 1939, **23**, 188

Bak F, Finster K, Rothfuss F, Formation of dimethylsulfide and methanethiol from methoxylated aromatic compunds and inorganic sulfide by newly isolate anaerobic bacteria, Arch Microbiol, 1992, **157**, 529

Berger RG, Genetic engineering III: Food flavors. In: Encyclopedia Food Sci Technol, Hui YH (ed) Wiley New York 1991, 1313

Berger RG, Neuhäuser K, Drawert F, High productivity fermentation of volatile flavours using a strain of *Ischnoderma benzoinum*, Biotechnol Bioeng, 1987, **30**, 987

Birkinshaw JH, Morgan EN, Volatile metabolic products of species of *Endoconidiophora*, J Biochem, 1950, **47**, 55

Birkinshaw JW, Kennedy Findlay WP, Metabolic products of *Lentinus lepideus* fr, J Biochem, 1940, **34**, 82

Bjurman J, Kristensson J, Volatile production by *Aspergillus versicolor* as a possible cause of odor in houses affected by fungi, Mycopathologia, 1992, **118**, 173

Boland W, Feng Z, Donath J, Oxidative bond cleavage; A general pathway to irregular terpenoids, In: Progress in Flavour Precursor Studies, Schreier P, Winterhalter P (eds) Allured Carol Stream IL 1993, 123

Boerjesson T, Stoellmann U, Schnuerer J, Volatile metabolites produced by six fungal species compared with other indicators of fungal growth on cereal grains, Appl Environ Microbiol, 1992, **58**, 2599

Boerjesson TS, Stollman UM, Schnurer JL, Off-odorous compunds produced by molds on oatmeal agar - identification and relation to other growth chracteristics, J Agric Food Chem, 1993, **41**, 2104

Buchbauer G, Jirovetz L, Wasicky M, Nikiforov A, Zum Aroma von Speisepilzen, Kopf-raum-Analyse mittels GC/FID und GC/FTIR/MS, Z Lebensm Unters Forsch, 1993, **197**, 429

Cailleux A, Bouchara JP, Daniel V, Chabasse D, Allain P, Gas chromatography-mass spectrometry analysis of volatile organic compounds produced by some micromycetes, Chromatographia, 1992, **34**, 613

Casey J, Dobb R, Microbial routes to aromatic aldehydes, Enzyme Microbial Technol, 1992, **14**, 739

Cerniglia CE, Sutherland JB, Crow SA, Fungal metabolism of aromatic hydrocarbons In: Microb Degrad Nat Prod, Winkelmann G (ed) VCH Weinheim 1992, 193

Chalier P, Crouzet J, Production of lactones by *Penicillium-roqueforti*, Biotechnol Let, 1992, **14**, 275

Cho KS, Hirai M, Shoda M, Enhanced removability of odorous sulfur-containing gases by mixed cultures of purified bacteria from peat biofilters, J Ferment Bioengin, 1992, **73**, 219

Colanbeen M, Neukermans G, Odour abatement of ventilation air by biofilters and bioscrubbers, Landbouwtijdschrift - Revue de L Agriculture, 1992, **45**, 909

Collett O, Aromatic compounds as growth substrates for isolates of the brown-rot fungus *Lentinus lepideus* (Fr.ex Fr.) Fr, Mater Org, 1992, **27**, 67

Collins RP, Knaak LE, Soboslai JW, Production of geosmin and 2-exo-hydroxy-2-methylbornane by *Streptomyces odorifer*, Lloydia, 1971, **33**, 199

Collins RP, Morgan ME, Identity of fruit-like aroma substances synthesized ba endoconi-dial-forming fungi, Phytopathology, 1962, **52**, 407

Collins RP, Terpenes and odoriferous materials from microorganisms, Lloydia, 1976, **39**, 20

Cormier F, Raymond Y, Champagne CP, Morin A, Analysis of odor-active volatiles from *Pseudomonas fragi* grown in milk, J Agric Food Chem, 1991, **39**, 159

Davies R, Falkiner EA, Wilkinson JF, Peel JL, Ester formation by yeast: 1. Ethyl acetate formation by *Hansenula* species, Biochem J, 1951, **49**, 58

Dembitsky VM, Rezanka T, Shubina EE, Unusual hydroxy fatty acids from some higher fungi, Phytochemistry, 1993, **34**, 1057

Deshusses MA, Hamer G, The removal of volatile ketone mixtures from air in biofilters, Biprocess Engin, 1993, **9**, 141

Drawert F, Barton H, Biosynthesis of flavor compounds by microorganisms. 3. Production of monoterpenes by the yeast *Kluyveromyces lactis,* J Agric Food Chem, 1978, **20**, 765

Dugelay I, Gunata Z, Sapis JC, Baumes R, Bayonove C, Role of cinnamoyl esterase acti-vities from enzyme preparations on the formation of volatile phenols during winemaking, J Agric Food Chem, 1993, **41**, 2092

Dwivedi BK Kinsella JE, Continuous production of blue-type cheese flavor by submerged fermentation of *Penicillium roqueforti*, J Food Sci, 1974, **39**, 620

Ercoli B, Fuganti C, Grasselli P, Servi S, Allegrone G, Barbeni M, Pisciotta A, Ste-reochemistry of the biogeneration of C-10 and C-12 gamma lactones in *Yarrowia lipolyti-ca* and *Pichia ohmeri,* Biotechnol Let, 1992, **14**, 665

Fiechter A, Function and synthesis of enzymes involved in lignin degradation, J Biotechnology, 1993, **30**, 49

Fritzsche Dodge & Olcott, US 4560656, 1985

Fuganti C, Fronza G, Grasselli P, Servi S, Allegrone G, Barbeni M, Cisero M, Stereochemistry of the microbil a generation of γ- and δ-lactones In: Progress in Flavour Precursor Studies, Schreier P, Winterhalter P (eds) Allured Carol Stream IL 1993, 451

Fukui S, Yagi T, Manufacture of methyl ketones and/or secondary alcohols with *Aureobasidium* and use of the culture media as flavors, JP 03247291 A2 911105, 1991

Gallois A, Gross B, Langlois D, Spinnler HE, Brunerie P, Influence of culture conditions on production of flavour compounds by 29 ligninolytic basidiomycetes, Mycol Res, 1990, **94**, 494

Gatfield IL, Güntert M, Sommer H Werkhoff P, Some aspects of the microbiological production of flavor-active lactones with particular reference to γ-decalactone, Chem Mikrobiol Technol Lebensm, 1993, **15**, 165

Gehring RF, Knight SG, Formation of ketones from fatty acids by spores of *Penicillium roqueforti,* Nature, 1958, **182**, 1237

Gerber NN, Geosmin from microorganisms is trans-1,10-dimethyl-*trans*-9-decalol, Tetrahedron Let, 1968, **25**, 2971

Gerber NN, Sesquiterpenoids from actionomycetes: cadin-4-ene-1-ol, Phytochemistry, 1971, **10**, 185

German JB, Berger RG, Drawert F, Generation of fresh fish flavour: Rainbow trout *Salmo gairdneri* gill homogenate as a model system, Chem Mikrobiol Technol Lebensm, 1991, **13**, 19

Gross B, Gallois A, Spinnler HE, Langlois D, Volatile compounds produced by the ligninolytic fungus *Phlebia radiata* Fr (Basidiomycetes) and influence of the strain specificity on the odorous profile, J Biotechnology, 1989, **10**, 303

Haarmann & Reimer, DE 3920039, 1991; EP 405197, 1990

Hädrich-Meyer S, Berger RG, Localization of lipolytic and esterolytic activities of *Tyromyces sambuceus,* a 4-decanolide producing basidiomycete, Appl Microbiol Biotechnol, 1994, **41**, 210

Hanssen HP, Abraham WR, Odoriferous compounds from liquid cultures of *Gloeophyllum odoratum* and *Lentinellus cochleatus* (Basidiomycotina), Flavour Fragrance J, 1987, **2**, 171

Hasegawa, JP 267284, 1990

Hattori S, Yamaguchi Y, Kanisawa T, Preliminary study on the microbiological formation of fruit flavors, Proc. IV Int. Congress Food Sci Technol, 1974, **1**, 143

Janssens L, De Pooter HL, De Mey L, Vandamme EJ, Schamp NM, Fruity flavours by fermentation, Med. Fac Landbouww Rijksuniv Gent, 1988, **53**, 2971

Jollivet N, Belin JM, Comparison of volatile flavor compounds produced by ten strains of *Penicillium camemberti* Thom, J Dairy Sci, 1993, **76**, 1837

Jollivet N, Bezenber MC, Vayssier Y, Belin JM, Procuction of volatile compounds in liquid cultures by six strains of coryneform bacteria, Appl Microbiol Biotechnol, 1992, **36**, 790

Kajiwara T, Matsui K, Hatanaka A, Tomoi T, Fujimura T, Kawai T, Distribution of an enzyme system producing seaweed flavor - conversion of fatty acids to long-chain aldehydes in seaweeds, J Appl Phycol, 1993, **5**, 225

Kaneshiro T, Vesonder RF, Peterson RE, Bagby MO, 2-hydroxyhexadecanoic and 8,9,13-trihydroxydocosanoic acid accumulation by yeasts treated with fumonisin-b-1, Lipids, 1993, **28**, 397

Kawabe T, Morita H, Volatile components in culture fluid of *Polyporus tuberaster*, J Agric Food Chem, 1993, **41**, 637

Kempler GM, Production of flavor compunds by microorganisms, Adv Appl Microbiol, 1983, **29**, 29

Kim CH, Rhee SK, Enzymatic conversion of ethanol into acetaldehyde in a gas-solid bioreactor, Biotechnol Let, 1992, **14**, 1059

Kim KH, Lee HJ, Optimum conditions for the formation of acetoin as a precursor of tetramethylpyrazine during the citrate fermentation by *Lactococcus lactis* subsp. *lactis* biovar *diacetilactis* FC 1, J Microbiol Biotechnol, 1991, **1**, 202 and 281

Klingenberg A, Sprecher E, Production of monoterpenes in liquid cultures by the yeast *Ambrosiozyma monospora*, Planta Medica, 1985, 264

Kobayashi H, Kohya S, Kawashima K, Kim WS, Tanaka H, Motoki M, Kasamo K, Kusakabe I, Deodorization of soybean protein with microorganisms, Biosci Biotech Biochem, 1992, **56**, 530

Koide K, Yanagisawa I, Fujita T, Studies on the kerosene-like odor production in 'surimi-based products', Nippon Suisan Gakkaishi, 1992b, **58**, 95

Koide K, Yanagisawa I, Ozawa A, Satake M, Fujita T, Studies on the kerosene-like off-flavour producing yeast in surimi-based products, Nippon Suisan Gakkaishi, 1992a, **58**, 1925

Krings U, PhD Thesis Universität Hannover, 1994

Labuda IM, Keon KA, Goers SK, Microbial bioconversion process for the production of vanillin In: Progress in Flavour Precursor Studies, Schreier P, Winterhalter P (eds) Allured Carol Stream IL 1993, 477

Laatsch H, Matthies L, The characteristic odor of *Coprinus picaceus* - A rapid enrichment procedure for apolar, volatile indoles, Mycologia, 1992, **84**, 264

Lanza E, Ko KH, Palmer JK, Aroma production by cultures of *Ceratocystic moniliformis*, J Agric Food Chem, 1976, **24**, 1247

Lanza E, Palmer JK, Biosynthesis of monoterpenes by *Ceratocystis moniliformis*, Phytochemistry, 1977, **16**, 1555

Latrasse A, Dameron P, Staron T, Production of a fruity aroma by *Geotrichum candidum* (Staron), Sci Aliments, 1987, **7**, 637

Latrasse A, Guichard E, Fournier N, Le Quéré JL, Dufossé L, Spinnler HE, Biosynthesis, chirality and flavour properties of the lactones formed by *Fusarium poae* In: Trends in Flavour Research, Maarse H, van der Heij (eds) Elsevier Amsterdam 1994, 493

Leete E, Bjorklund JA, Biosynthesis of 3-isopropyl-2-methoxypyrazine and other alkylpyrazines: Widely distributed flavour compounds In: Bioformation of Flavours, Patterson RLS, Charlwood BV, MacLeod G, Williams AA (eds) Royal Soc Chem Cambridge 1992, 75

Lemenager Y, Pissot de Leffernberg C, Helaine D, Marc Y, Bioconversion of lipids in emulsions for the preparation of aroma components, FR 2664288/A1/900110, 1992

Mann S, Über den Geruchssstoff von Pseudomonas aeruginosa, Arch Mikrobiol, 1966, 54, 184

Mattheis JP, Roberts RG, Identification of geosmin as a volatile metabolite of Penicillium expansum, Appl Environm Microbiol, 1992, 58, 3170

Mau JL, Beelman RB, Ziegler GR, Effect of 10-oxo-trans-8-decenoic acid on growth of Agaricus bisporus, Phytochemistry, 1992, 31, 4059

Mau JL, Beelman RB, Ziegler GR, Factors affecting 1-octen-3-ol in mushrooms at harvest and during postharvest storage, J Food Sci, 1993, 2, 331

Miersch O, Günther T, Fritsche W, Sembdner G, Jasmonates from different fungal species, Nat Product Let, 1993, 2, 293

Nago H, Matsumoto M, Nakai S, 2-Deceno-δ-lactone-producing, fungi, strains of Fusarium solani, Isolated by using a medium containing decano-δ-lactone as the sole carbon source, Biosci Biotech Biochem, 1993, 57, 2107

Okui S, Uchiyama M, Mizugaki M, Metabolism of hydroxy fatty acids. Intermediates of the oxidative breakdown of ricinoleic acid by genus Candida, J Biochem, 1963, 54, 536

Omelianski VL, Aroma-producing microörganisms, J Bacteriol, 1923, 8, 393

Onodera R, Ueda H, Nagasawa T, Oku K, Chaen S, Mieno M, Kuto H, In vitro metabolism of tryptophan by ruminal protozoa and bacteria. The production of indole and skatole and their effects on protozoal survival and VFA production, Anim Sci Technol, 1992, 63, 23

Page GV, Farbood MI, A method for the preparation of 'natural' methyl anthranilate, WO 89//00203, 1989

Park SJ, Hirai M, Shoda M, Treatment of exhaust gases from a night soil treatment plant by a combined deodorization system of activated carbon fabric reactor and peat biofilter inoculated with Thiobacillus thioparus DW44, J Fermentation Bioengin, 1993, 76, 423

Pernod-Ricard, FR 9005003, 1991

Pollak F, PhD Thesis, Universität Hannover, 1994

Pyysalo H, Identification of volatile compunds in seven edible fresh mushrooms, Acta Chemica Scandinavica B, 1976, 30, 235

Quest, EP 91202920, 1991

Raymond Y, Morin A, Champagne CP, Cormier F, Enhancement of fruity aroma production of Pseudomonas fragi grown on skim milk, whey and whey permeate supplemented with C_3-C_7 fatty acids, Appl Microbiol Biotechnol, 1991, 34, 524

Rezanka T, Líbalová D, Votruba J, Víden I, Identification of odorous compounds from Streptomyces avermitilis, Biotechnol Let, 1994, 16, 75

Schindler, J, Terpenoids by microbial fermentation, Ind Eng Chem Prod Res Dev, 1982, 21, 537

Serrano-Carreon L, Hathout Y, Bensoussan M, Belin JM, Production of 6-pentyl-alpha-pyrone by Trichoderma harzianum from 18n fatty acid methyl esters, Biotechnol Let, 1992, 14, 1019

Shareefdeen Z, Baltzis BC, Oh YS, Bartha R, Biofiltration of methanol vapor, Biotechnol Bioengin, 1993, **41**, 512

Spinnler HE, Djian A, Bioconversion of amino acids into flavouring alcohols and esters by *Erwinia carotovora* subsp. *atroseptica*, Appl Microbiol Biotechnol, 1991, **35**, 264

Spinnler HE, Grosjean O, Bouvier I, Effect of culture parameters on the production of styrene (vinyl benzene) and 1-octene-3-ol by *Penicillium caseicolum*, J Dairy Res, 1992, **59**, 533

Sprecher E, Kubeczka KH, Ratschko M, Flüchtige Terpene in Pilzen, Archiv Pharmazie, 1975, **308**, 843

Stutz HK, Silverman GJ, Angelini P, Levin RE, Bacteria and volatile compunds associated with ground beef spoilage, J Food Sci, 1991, **56**, 1147

Teranishi R, Takeoka GR, Güntert DM (eds) 1992, Flavor Precursors, ACS symp ser 490, ACS Wash DC

Tiefel P, Berger RG, Seasonal variation of the concentrations of maltol and maltol glucoside in leaves of *Cercidiphyllum japonicum*, J Sci Food Agric, 1993b, **63**, 59

Tiefel P, Berger RG, Volatiles in precusor fed cultures of *Basidiomycetes* In: Progress in Flavor Precusor Studies, Schreier P, Winterhalter P (eds) Allured Carol Stream IL 1993a, 439

Tiefel P, PhD Thesis TU München 1994

Tressl R, Albrecht W, Kersten E, Nittka C Rewicki D, Enzymatic and thermal conversion of (^{2}H)- and (^{13}C)-labelled precursors to flavour compounds, In: Progress in Flavour Precursor Studies, Schreier P, Winterhalter P (eds) Allured Carol Stream IL 1993, 1

Tsuchiya Y, Watanabe MF, Watanabe M, Volatile organic sulfur compounds associated with blue-green algae from inland waters of Japan, Water Sci Technol, 1992, **25**, 123

Tuor UM, Shoemaker HE, Leisola MSA, Schmidt HWH, Isolation and characterization of substituted 4-hydroxy-cyclohex-2-enones as metabolites of 3,4-dimethoxybenzyl alcohol and its methyl ether in ligninolytic cultures of *Phanerochaete chrysosporium*, Appl Microbiol Biotechnol, 1993, **38**, 674

van der Schaft PH, ter Burg N, van den Bosch S, Cohen AM, Fed-batch production of 2-heptanone by *Fusarium poae*, Appl Microbiol Biotechnol, 1992a, **36**, 709

van der Schaft PH, ter Burg N, van den Bosch S Cohen, AM, Microbial production of natural δ-decalactone and δ-dodecalactone from the corresponding α,β-unsaturated lactones in *Massoi* bark oil, Appl Microbiol Biotechnol 1992b, **36**, 712

Vick BA, Zimmermann DC, Biosynthesis of jasmonic acid by several plant species, Plant Physiol, 1984, **75**, 458

Wainwright M, An Introduction to Fungal Biotechnology, Wiley Chichester 1992

Williams TO, Miller FC, Facility managment - odor control using biofilters, Biocycle, 1992, **33**, 72

Wnorowski AU, Scott WE, Incidence of off-flavors in South African surface waters, Water Sci Technol, 1992, **25**, 225

Zeng XN, Leyden JJ, Brand JG, Spielman AI, McGinley KJ, Preti, G, An investigation of human apocrine gland secretion for axillary odor precursors, J Chem Ecol, 1992, **18**, 1039

Zeringue HJ, Bhatnagar D, Cleveland TE, C15H24 Volatile compounds unique to aflatoxigenic strains of *Aspergillus flavus*, Appl Environm Microbiol, 1993, **59**, 2264

Chapter 6

Abraham WR, Ernst L, Stumpf B, Arfmann HA, Microbial hydroxylations of bicyclic and tricyclic sesquiterpenes, J Ess Oil Res, 1989, 1, 19

Abraham WR, Kieslich K, Stumpf B, Ernst L, Microbial oxidation of tricyclic sesquiter-penoids containing a dimethylcyclopropane ring, Phytochemistry, 1992, 31, 3749

Abraham WR, Stumpf B, Biotransformation of humulene by fungi and enantioselectivity of the strains used, Z Naturforsch, 1987, 42c, 79

Arfmann HA, Abraham WR, Kieslich K, Microbial ω-hydroxylation of trans-nerolidol and structurally related sesquiterpenoids, Biocatalysis, 1988, 2, 59

Arfmann HA, Abraham WR, Microbial reduction of aromatic carboxylic acids, Z Natur-forsch, 1993, 48c, 52

Boopathy R, Bokang H, Daniels L, Biotransformation of furfural and 5-hydroxymethyl furfural by enteric bacteria, J Ind Microbiol , 1993, 11, 147

Busmann D, PhD Thesis Universität Hannover 1994

Cardillo R, Nasini G, Depava OV, Secondary mould metabolites 36. Isolation of illudin-m, illudinine and shisool from the mushroom *Clitocybe phosphorea* - Biotransformation of perillyl alcohol and aldehyde, Phytochemistry, 1992, 31, 2013

Dhavalikar RS. Bhattacharyya P.K. Microbiological transformations of terpenes: Part VIII - Fermentation of limonene by a soil pseudomonad, Indian J Biochem, 1966, 3, 144

Diaz DME, Villa P, Guerra M, Rodriguez E. Redondo D, Martinez A, Conversion of furfural into furfuryl alcohol by *Saccharomyces cervisiae*, Acta Biotechnol, 1992, 12, 351

Doi M, Matsui M, Kanayama T, Shuto Y, Kinoshita Y, Asymmetric reduction of aceto-phenone by *Aspergillus* Species and their possible contribution to katsuobushi flavor, Biosci Biotechnol Biochem, 1992, 56, 958

Hanson JR, Nasir H, Biotransformation of the sesquiterpenoid, cedrol, by *Cephalosporium aphidicola*, Phytochemistry, 1993, Vol 33, 4, 835

Horiuchi K, Morimoto KI, Ohta T, Suemitsu R, Biotransformation of benzyl alcohol by *Pseudomonas cepacia*, Biosci Biotech Biochem, 1993, 57, 1346

Ishida T, , Asakawa Y, Takemoto T, Aratani T, Terpenoids Biotransformation in mam-mals III: Biotransformation of α-pinene, β-pinene, pinane, 3-carene, carane, myrcene, and p-cymene in rabbits, J Pharmac Sci, 1981, 70, 406

Kieslich K, Abraham WR, Stumpf B, Washausen P, Microbial transformation of terpe-noids, In: Topics in Flavour Research, Berger RG, Nitz S, Schreier P (eds) Eichhorn Han-genham 1985, 405

Koritala S, Bagby MO, Microbial conversion of linoleic and linolenic acids to unsaturated hydroxy fatty acids, JAOCS, 1992, 69, 575

Krasnobajew V, Helmlinger D, Fermentation of fragrances: biotransformation of β-ionone by *Lasiodiplodia theobromae*, Helv Chim Acta, 1982, 65, 1590

Lamare V, Furstoss R, Bioconversion of sesquiterpenes, Tetrahedron, 1990, 46, 4109

Lanser AC, Conversion of oleic acid to 10-ketostearic acid by a *Staphylococcus* species, JAOCS, 1993, 70, 543

Lanser AC, Plattner RD, Bagby MO, Production of 15-, 16-and 17-hydroxy-9-octadecanoic acids by bioconversion of oleic acid with *Bacillus pumilus*, JAOCS, 1992, **69**, 363

Maatooq G, Elsharkawy S, Afifi MS, Rosazza JPN, Microbial transformation of cedrol, J Nat Prod 1993, **56**, 1039

Madyastha KM, Gururaja TL, Transformations of acyclic isoprenoids by *Aspergillus niger* - selective oxidation of omega-methyl and remote double bonds, Appl Microbiol Biotechnol, 1993, **38**, 738

Manzoni M, Molinari F, Tirelli A, Aragozzini F, Phenylacetaldehyde by acetic acid bacteria oxidation of 2-phenylethanol, Biotechnol Let, 1993, **15**, 341

Mauersberger S, Drechsler H, Oehme G, Müller HG, Substrate specificity and stereoselectivity of fatty alcohol oxidase from the yeast *Candida maltosa*, Appl Microbiol Biotechnol, 1992, **37**, 66

Mikami Y, Fukunaga Y, Arita M, Kisaki T, Microbial transformation of β-ionone and β-methylionone, Appl Environm Microbiol, 1981, **41**, 610

Noma Y, Yamasaki S, Asakawa Y, Biotransformation of limonene and related componds by *Aspergillus cellulosae*, Phytochemistry, 1992, **31**, 2725

Ohta A, Shimada M, Aromatic hydroxylation of methyl cinnamate to methyl 4-hydroxycinnamate catalyzed by the cell-free extracts of a brown-rot fungus *Lentinus lepideus*, Mokuzai Gakkaishi, 1991, **37**, 748

Patel RN, McNamee CG, Banerjee A, Howell JM, Robison RS, Szarka LJ, Stereoselective reduction of beta-keto esters by *Geotrichum candidum*, Enzyme Microbial Technol, 1992, **14**, 731

Phillips Petroleum , US 5071762, 1991

Repp HD, Stottmeister U, Dörre M, Weber L, Haufe G, Microbial oxidation of α-pinene, Biocatalysis 1990, **4**, 75

Rosazza JPN, Goswami A, Liu WG, Sariaslani FS, Steffens JJ, Steffek RP, Beale JM, Chapman JR, Reeg S, Microbial transformations of terpenes: studies with 1,4-cineole, J Ind Microbiol 1988, **29**, 181

Sandey H, Willetts A, Biotransformation of cycloalkanones - controlled oxidative and reductive bioconversions by an *Acinetobacter* species, Biotechnol Let, 1992, **14**, 1119

Stumpf B, Kieslich K, Appl Microbiol Biotechnol 1991, **34**, 598

Stumpf B, Wray V, Kieslich K, Oxidation of carenes to chaminic acids by *Mycobacterium smegmatis* DSM 43061, Appl Microbiol Biotechnol, 1990, **33**, 251

Suhara Y, Itoh S, Ogawa M, Yokose K, Sawada T, Sano T, Ninomiya R, Maruyama HB, Regio-selective 10-hydroxylation of patchoulol, a sesquiterpene, by *Pithomyces* Species, Appl Environm Microbiol, 1981, **42**, 2, 187

Takigawa H, Kubota H, Sonohara H, Okuda M, Tanaka S, Fujikura Y, Ito S, Novel allylic oxidation of alpha-cedrene to sec-cedrenol by a *Rhodococcus* strain, Appl Environm Microbiol, 1993, **59**, 1336

van der Schaft P, van Geel I, de Jong G, ter Burg N, Microbial production of natural furfurylthiol, In: Trends in Flavour Research, Maarse H, van der Heij DG (eds) Elsevier Amsterdam 1994, 437

Venosa AD, Haines JR, Nisamaneepong W, Govind R, Pradhan S, Siddique B, Efficacy of commercial products in enhancing oil biodegradation in closed laboratory reactors, J Ind Microbiol, 1992, **10**, 13

Villa GP, Bartroli R, Lopez R, Guerra M, Enrique M, Penas M, Rodriquez E, Redondo D, Iglesias I, Diaz M, Microbial transformation of furfural to furfuryl alcohol by *Saccharomyces cerevisiae*, Acta Biotechnol, 1992, **12**, 509

Vischer E, Wettstein A, Mikrobiologische Reaktionen, Experientia, 1953, **9**, 371

Chapter 7

Adedeji J, Hartman TG, Lech J, Ho CT, Chracterization of glycosidically bound aroma compunds in the African mango (*Mangifera indica* L), J Agric Food Chem, 1992, **40**, 659

Anderson DA, A matter of taste, Chemtech, 1993, **23**, 45

Berger RG, Genetic Engineering III: Food Flavors, In: Encyclopedia Food Science Technology, Hui YH (ed), Wiley New York 1991, 1313

Braun SD, Olsen NF, Microencapsulation of cell-free extracts to demonstrate the feasability of heterogenous enzyme systems and cofactor recycling for development of flavor in cheese, J Dairy Sci, 1986, **69**, 1202

Calis I, Saracoglu I, Basaran AA, Sticher O, 2 phenethyl alcohol glycosides from *Scutellaria orientalis* subsp *pinnatifida*, Phytochemistry, 1993, **32**, 1621

Cao SG, Liu ZB, Feng Y, Ma L, Ding ZT, Cheng YH, Esterification and transesterification with immobilized lipase in organic solvent, Appl Biochem Biotechnol, 1992, **32**, 1

Ceynowa J, Sionkowska I, Enantioselective esterification of butanol-2 in an enzyme membrane reactor, Acta Biotechnol, 1993, **13**, 177

Challier P, Koulibaly AA, Fontvieille MJ, Crouzet J, Fruit glycosides as aroma precursors, Int Fruchtsaft-Union, Wiss-Tech Komm, 1990, **21**, 349

Chartrain M, Katz L, Marcin C, Thien M, Smith S, Fisher E, Goklen K, Salmon P, Brix T, Price K, Greasham R, Purification and characterization of a novel bioconverting lipase from *Pseudomonas aeruginosa* MB 5001, Enzyme Microbial Technol, 1993, **15**, 575

Chattopadhyay S, Mamdapur VR, Enzymatic esterification of 3-hydroxybutyric acid, Biotechnol Let, 1993, **15**, 245

Cheetham PSJ, Novel specific pathways for flavour production, In: Bioformation of Flavours, Patterson RLS, Charlwood BV, MacLeod G, Williams AA (eds) Royal Soc Chem Cambridge 1992, 96

Chen JP, Wu S-H, Wang, K-T, Double enantioselective esterification of racemic acids and alcohols by lipase from *Candida cylindracea*, Biotechnol Let 1993, **15**, 181

Chen JP, Yang BK, Enhancement of release of short-chain fatty acids from milk fat with immobilized microbial lipase, J Food Sci, 1992, **57**, 781

Creagh AL, Prausnitz JM, Blanch HW, Structural and catalytic properties of enzymes in reverse micelles, Enzyme Microbial Technol, 1993, **15**, 383

Daiwa Kasei KK, JP 03240432 A2 011025, 1991

De Tommasi N, Aquino R, Desimone F, Pizza C, Plant metabolites - new sesquiterpene and ionone glycosides from *Eriobotrya japonica*, J Nat Prod, 1992, **55**, 1025

Di Cesare LF, Nani R, Forni E, Scotto P, Influence of pectolytic enzymes on the flow rate and flavoring of peach juice, Fluess Obst, 1993, **60**, 254

Dugelay I, Gunata Z, Sapis JC, Baumes R, Bayonove C, Role of cinnamoyl esterase activities from enzyme preparations on the formation of volatile phenols during winemaking, J Agric Food Chem, 1993, **41**, 2092

Dupin I, Gunata Z, Sapis JC, Bayonove C, Mbairaroua O, Tapiero C, Production of beta-apiosidase by *Aspergillus niger* - partial purification, properties, and effect on terpenyl apiosylglucosides from grape, J Agric Food Chem, 1992, **40**, 1886

Ebara SJ, JP 05268905 A2 931019, 1993

Engel KH, Lipases: Useful biocatalysts for enantioselective reactions of chiral flavor compounds, In: Flavor Precursors, Teranishi R, Takeoka GR, Güntert M (eds) ACS symp ser 490, ACS Wash DC 1992, 21

Faber K, 1992 Biotransformations in Organic Chemistry, Springer Berlin

Francis MJO, Allcock C, Geraniol β-D-glucoside, occurrence and synthesis in rose flowers, Phytochemistry, 1969, **8**, 1339

Francis IL, Sefton MA, Williams PJ, Sensory Descriptive analysis of the aroma of hydrolysed precursor fractions from *Semillon*, *Chardonnay* and *Sauvignon blanc* grape juices, J Sci Food Agric, 1992, **59**, 511

Gargouri Y, Cudrey C, Mejdoub H, Verger R, Inactivation of human pancreatic lipase by 5-dodecyldithio-2-nitrobenzoic acid, Eur J Biochem, 1992, **204**, 1063

Gist-Brocades NV, EP 416713 Al 910313, 1991

Givaudan-Roure, WO 94/07887, 1994

Guo W, Sakata K, Watanabe N, Nakajima R, Yagi A, Ina K, Luo S, Geranyl 6-O-β-D-xylopyranosyl-β-D-glucopyranoside isolated as an aroma precursor from tea leaves for oolong tea, Phytochemistry, 1993, **33**, 1373

Hädrich-Meyer S, Berger RG, Localization of lipolytic and esterolytic activities of *Tyromyces sambuceus*, a 4-decanolide producing basidiomycete, Appl Microbiol Biotechnol, 1994, 41, 210

Herderich M, Neubert C, Winterhalter P, Schreier P, Skouroumounis GK, Identification of C13-norisoprenoid flavor precursors in starfruit (*Averrhoa carambola* L), Flavour Fragrance J, 1992, **7**, 179

Horowitz JB, Vilker VL, Development of bacterial cytochrome p-450cam (*Cytochrome m*) production, Biotechnol Bioengin, 1993, **41**, 411

Humpf HU, Schreier P, 3-hydroxy-5,6-epoxy-beta-ionol beta-D-glucopyranoside and 3-hydroxy-7,8-dihydro-beta-ionol beta-D-glucopyranoside - new C13 norisoprenoid glucoconjugates from sloe tree (*Prunus spinosa* L) leaves, J Agric Food Chem, 1992, **40**, 1898

Humpf HU, Wintoch H, Schreier P, 3,4-dihydroxy-7,8-dihydro-beta-ionol beta-d-glucopyranoside - natural precursor of isomeric vitispiranes from gooseberry (*Ribes uva crispa* L) and whitebeam (*Sorbus aria*) leaves, J Agric Food Chem, 1992, **40**, 2060

Hwang SO, Trantolo DJ, Wise DL, Gas phase acetaldehyde production in a continuous bioreactor, Biotechnol Bioengin, 1993, **42**, 667

Ishikawa S, Ito Y, Nakaya H, Development of new food materials by using enzymes and membrane reactors, Aomori-ken Suisanbutsu Kako Kenkyusho Kenkyu Hokoku, 1993, 39, after CA **120**: 105306d

Itozawa T, Kise H, Highly efficient coenzyme regeneration system for reduction of cyclohexanone by horse liver alcohol dehydrogenase in organic solvent, Biotechnol Let, 1993, **15**, 843

Izumi T, Nakamura T, Eda Y, Enzymatic synthesis of (R)-2-cyclohexen-1-ol and (2)-2-cyclohexen-1-ol, J Chem Technol Biotechnol, 1993, **57**, 175

Karp F, Croteau R, Role of hydroxylases in monoterpene biosynthesis, In: Bioflavour '87, Schreier P (ed) deGruyter Berlin 1988, 173

Kataoka M, Nomura Y, Shimizu S, Yamada H, Enzymes involved in the NADPH regeneration system coupled with asymmetric reduction of carbonyl compounds in microorganisms, Biosci Biotech Biochem, 1992, **56**, 820

Kazandjian RZ, Dordick JS, Klibanov M, Enzymatic analyses in organic solvents, Biotechnol Bioengin, 1986, **28**, 417

Konishi Y, Kitazato S, Nakatani N, Partial purification and characterization of acid and neutral alpha-glucosidases from preclimacteric banana pulp tissues, Biosci Biotech Biochem, 1992, **56**, 2046

Koulibaly A, Sakho M, Crouzet J, Variability of free and bound volatile terpenic compunds in mango, Food Sci Technol, 1992, **25**, 374

Langrand G, Rondot N, Triantaphylides C, Baratti J, Short chain flavour esters synthesis by microbial lipases, Bitechnol Let, 1990, **12**, 8, 581

Lee KH, Lee PM, Siaw YS, Morihara K, Effects of methanol on the synthesis of aspartame precursor catalysed by *Pseudomonas aeruginosa* elastase, Biotechnol Let, 1992, **14**, 779 and 1159

Lindsay RC, Rippe JK, Enzymic generation of methanethiol to assist in the flavor development of cheddar cheese and other foods, In: Biogeneration of Aromas, Parliment TH, Croteau R (eds) ACS symp ser 317, ACS Wash DC 1986, 286

Lopez V, Lindsay RC, Metabolic conjugates as precursors for characterizing flavor compounds in ruminant milks, J Agric Food Chem, 1993, **41**, 446

Loughrin JH, Hamilton-Kemp TR, Burton HR, Andersen RA, Hildebrand DF, Glycosidically bound volatile components of *Nicotiana sylvestris* and *N. suaveolens* flowers, Phytochemistry, 1992, **31**, 1537

Lutz D, Huffer M, Gerlach D, Schreier P, Carboxylester-lipase-mediated reactions, In: Flavor Precursors, Teranishi R, Takeoka GR, Güntert M (eds) ACS symp ser 490, ACS Wash DC 1992, 32

Marlatt C, Ho CT, Chien M, Studies of aroma constituents bound as glycosides in tomato, J Agric Food Chem, 1992, **40**, 249

Marinos VA, Tate ME, Williams PJ, Lignan and phenylpropanoid glycerol glucosides in wine, Phytochemistry, 1992, **31**, 4307

Matsui I, Yoneda S, Ishikawa K, Miyairi S, Fukui S, Umeyama H, Konda K, Roles of the aromatic residues conserved in the active center of *Saccharomycopsis* alpha-amylase for transglycosylation and hydrolysis activity, Biochemisty, 1994, **33**, 451

Minetoki T, Alcohol acetyltransferase of sake yeast, Nippon Jozo Kyokaishi, 1992, **87**, 334

Miyazawa T, Kurita S, Ueji S, Yamada T, Kuwata S, Resolution of racemic carboxylic acids via the lipase-catalyzed irreversible transesterification using vinyl esters - effects of alcohols as nucleophiles and organic solvents on enantioselectivity, Biotechnol Let, 1992, **14**, 941

Pabst A, Barron D, Etiévant P, Schreier P, Studies on the enzymatic hydrolysis of bound aroma constituents from raspberry fruit pulp, J Agric Food Chem, 1991, **39**, 173

Park SK, Morrison JC, Adams DO, Noble AC, Distribution of free and glycosidically bound monoterpenes in the skin and mesocarp of muscat of *Alexandria* grapes during development, J Agric Food Chem, 1991, **39**, 514

Park YK, Uekane RT, Pupin AMK, Conversion of sucrose to isomaltulose by microbial glycosyltransferase, Biotechnol Let, 1992, **14**, 547

Parvaresh F, Robert H, Thomas D, Legoy MD, Gas phase transesterification reactions catalyzed by lipolytic enzymes, Biotechnol Bioengin, 1992, **39**, 467

Pernod-Ricard, WO 9304597 Al 930318, 1993

Peters J, Zelinski T, Kula MR, Studies on the distribution and regulation of microbial keto ester reductases, Appl Microbiol Biotechnol, 1992, **38**, 334

Ranadive AS, Vanillin and related flavor compounds in vanillia extracts made from beans of various global origins, J Agric Food Chem, 1992, **40**, 922

Razungles A, Gunata Z, Pinatel S, Baumes R, Bayonove C, Quantitative studies on terpenes, norisoprenoides and their precursors in several varieties of grapes, Sci Aliments, 1993, **13**, 59

Roscher R, Winterhalter P, 1,1,6-trimethyl-1,2-dihydronaphthalene (TDN) formation in wine .2. application of multilayer coil countercurrent chromatography for the study of *Vitis vinifera* cv *Riesling* leaf glycosides, J Agric Food Chem, 1993, **41**, 1452

Salles C, Jallageas JC, Fournier F, Tabet JC, Crouzet JC, Apricot glycosidically bound volatile components, J Agric Food Chem, 1991, **39**, 1979

Schreier P, Winterhalter P (eds), 1993 Progress in Flavour Precursor Studies, Allured Carol Stream IL

Sefton MA, Winterhalter P, Williams PJ, Free and bound 6,9-dihydroxymegastigm-7-en-3-one in *Vitis vinifera* grapes and wine, Phytochemistry, 1992, **31**, 1813

Shimada Y, Koga C, Sugihara A, Nagao T, Takada N, Tsunasawa S, Purification and characterization of a novel solvent-tolerant lipase from *Fusarium heterosporum*, J Ferment Bioeng, 1993, **75**, 343

Sonnet PE, Lipase-catalyzed hydrolysis - effect of alcohol configuration on the stereobias for 2-methyloctanoic acid, J Agric Food Chem, 1993, **41**, 319

Stamatis H, Xenakis A, Kolisis FN, Enantiomeric selectivity of a lipase from *Penicillium simplicissimum* in the esterification of menthol in microemulsions, Biotechnol Let, 1993, **15**, 471

Stöcklein W, Sztajer H, Menge U, Schmid RD, Purification and properties of a lipase from *P. expansum* - enzyme isolation and characterization, Biochim Biophys Acta L 1993, **1168**, 181

Sugiyama M, Nagayama E, Kikuchi M, Studies on the constituents of *Osmanthus species* 14. lignan and phenylpropanoid glycosides from *Osmanthus asiaticus*, Phytochemistry, 1993, **33**, 1215

Takeoka GR, Flath RA, Buttery RG, Winterhalter P, Guentert M, Ramming DW, Tera-nishi R, Free and bound flavor constituents of white-fleshed nectarines, In: Flavor precur-sors , Teranishi R, Takeoka GR, Güntert M (eds) ACS symp ser, 490, ACS Wash DC 1992, 116

Tanaka S, Karwe MV, Ho CT, Glycoside as a flavor precursor during extrusion, In: Ther-mally Generated Flavors, Teranishi R, Takeoka GR, Güntert M (eds) ACS symp ser 543, ACS Wash DC 1994, 370

Tengerdy RP, Linden JC, Wu M, Murphy VG, Baintner F, Szakacs G, Plant processing by simultaneous lactic acid fermentation and enzyme hydrolysis, Appl Biochem Biotechnol, 1992, **34**, 309

Teranishi R, Takeoka GR, Güntert M (eds),1992, Flavor Precursors, ACS symp ser 490, ACS Wash DC

Triantafyllou AO, Adlercreutz P, Mattiasson B, Influence of the reaction medium on enzyme activity in bio-organic synthesis - behaviour of lipase from *Candida rugosa* in the presence of polar additives, Biotechnol Appl Biochem, 1993, **17**, 167

Unno T, Ide K, Yazaki T, Tanaka Y, Nakakuki T, Okada G, High recovery purification and some properties of a β-glucosidase from *Aspergillus niger*, Biosci Biotech Biochem, 1993, **57** 2172

Vasserot Y, Arnaud A, Galzy P, Evidence for muscat marc monoterpenol glucosides hy-drolysis by free or immobilized yeast beta-glucosidase, Bioresource Technol, 1993, **43**, 269

Vulfson EN, Enzymic synthesis of food ingredients in low-water media, Trends Food Sci Technol, 1993, **4**, 209

Watanabe N, Watanabe S, Nakajima R, Moon JH, Shimokihara K, Inagaki J, Etoh H, Asai T, Sakata K, Ina K, Formation of flower fragrance compounds from their precursors by enzymic action during flower opening, Biosci Biotech Biochem, 1993, **57**, 1101

Whitaker JR, Importance of enzymes to value-added quality of foods, Food Structure, 1992, **11**, 201

Wilcocks R, Ward OP, Collins S, Dewdney NJ, Hong YP, Prosen E, Acyloin formation by benzoylformate decarboxylase from *Pseudomonas putida*, Appl Environm Micorbiol, 1992, **58**, 1699

Williams PJ, Sefton MA, Marinos VA, Hydrolytic flavor release from nonvolatile precur-sors in fruits, wines and some other plant-derived foods In: Hopp R Mori K (eds) Recent Developments in Flavor and Fragrance Chemistry, VCH Weinheim 1993, 283

Wintoch H, Krammer G, Schreier P, Glycosidically bound aroma compunds from two strawberry fruit species, *Fragaria vesca f. semperflorens* and *Fragaria×ananassa, cv Korona*, Flavour Fragrance J, 1991, **6**, 209

Wintoch H, Morales A, Duque C, Schreier P, (R)-(-)-(E)-2,6 dimethyl-3,7-octadiene-2,6-diol 6-o-beta-D-glucopyranoside - natural precursor of hotrienol from lulo fruit (*Solanum vestissimum* D) peelings, J Agric Food Chem, 1993, **41**, 1311

Wong CH, Mazenod FP, Whitesides GM, Chemical and enzymatic syntheses of 6-deoxyhexoses. Conversion to 2,5-dimethyl-4-hydroxy-2,3 dihydrofuran-3-one (furaneol) and analogues, J Org Chem, 1983, **48**, 3493

Wu P, Kuo MC, Hartman TG, Rosen RT, Ho CT, Free and glycosidically bound aroma compounds in pineapple (*Ananas comosus* L Merr), J Agric Food Chem, 1991, **39**, 170

Yoshikawa K, Nagai M, Wakabayashi M, Arihara S, Aroma glycosides from *Hovenia-dulsis*, Phytochemistry, 1993, **34**, 1431

Chapter 8

Akita O, Breeding of sake yeast producing a large quantity of aroma, Nippon Jozo Kyo-kaishi, 1992, **87**, 621

Aoki T, Uchida K, Amino acids-uptake deficient mutants of *Zygosaccharomyces rouxii* with altered production of higher alcohols, Agric Biol Chem, 1991, **55**, 2893

Barbosa MDS, Beck MJ, Fein JE, Potts D, Ingram LO, Efficient fermentation of *Pinus* sp acid hydrolysates by an ethanologenic strain of *Escherichia coli*, Appl Environ Microbiol, 1992, **58**, 1382

Birkeland SE, Abrahamsen RK, Langsrud T, Accelerated cheese ripening - use of lac-mutants of *Lactococci*, J Dairy Res, 1992, **59**, 389

Bosetti A, Vanbeilen JB, Preusting H, Lageveen RG, Witholt B, Production of primary aliphatic alcohols with a recombinant *Pseudomonas* strain, encoding the alkane hydroxy-lase enzyme system, Enzyme Microbial Technol, 1992, **14**, 702

Breidt F, Fleming HP, Competitive growth of genetically marked malolactic-deficient *Lactobacillus plantarum* in cucumber fermentations, Appl Environm Microbiol, 1992, **58**, 3845

Casey R, Domoney C, Ealing P, North H, The biochemical genetics of lipoxygenases, JAOCS, 1989, **66**, 491

Chagnaud P, Chion CKNCK, Duran R, Naouri P, Arnaud A, Galzy P, Construction of a new shuttle vector for *Lactobacillus*, Can J Microbiol, 1992, **38**, 69

Chung GH, Leer YP, Jeohn GH, Yoo OJ, Rhee JS, Cloning and nucleotide sequence of thermostable lipase gene from *Ps. fluorescens* Sik W1, Agric Biol Chem, 1991, **55**, 2359

Compagno C, Tura A, Ranzi BM, Martegani E, Bioconversion of lactose whey to fructose diphosphate with recombinant *Saccharomyces cerevisiae* cells, Biotechnol Bioengin, 1993, **42**, 398

Dell KA, Frost JW, Identification and removal of impediments to biocatalytic synthesis of aromatics from D-glucose: rating-limiting enzymes in the common pathway of aromatic amino acid biosynthesis, JACS, 1993, **115**, 11581

Demirci A, Pometto AL, Enhanced production of D(-)-lactic acid by mutants of *Lac-tobacillus delbrueckii* ATCC-9649, J Ind Microbiol, 1992, **11**, 23

Devos WM, Mulders JWM, Siezen RJ, Hugenholtz J, Kuipers OP, Properties of nisin-z and distribution of its gene, nisZ, in *Lactococcus lactis*, Appl Environ Microbiol, 1993, **59**, 213

Elalami N, Boquien CY, Corrieu G, Batch cultures of recombinant *Lactococcus lactis* ssp *lactis* a stirred fermentor 2. plasmid transfer in mixed cultures, Appl Microbiol Biotech-nol, 1992a, **37**, 364

Elalami N, Boquien CY, Corrieu G, Batch cultures of recombinant *Lactococcus lactis* ssp *lactis* in a stirred fermentor 1. effect of plasmid content on bacterial growth and on genetic stability in pure cultures, Appl Microbiol Biotechnol, 1992b, **37**, 358

Favrebulle O, Weenink E, Vos T, Preusting H, Witholt B, Continuous bioconversion of n-octane to octanoic acid by recombinant *Escherichia coli* (alk+) growing in 2-liquid-phase chemostat, Biotechnol Bioengin, 1993, **41**, 263

Feller G, Thirg M, Arpigny JL, Gerday C, Cloning and expression in *E. coli* of three lipase-encoding genes from the psychrotrophic antarctic strain *Moraxella* TA 144, Gene, 1991, **102**, 111

Fukuda K, Muromachi A, Watanabe M, Asano K, Takasawa S, Mutants producing high concentrations of the flavor components active amylalcohol and normal propanol in *Saccharomyces cerevisiae*, J Fermentation Bioengin, 1993, **75**, 288

Fukuda K, Watanabe M, Asano K, Ouchi K, Takasawa S, Molecular breeding of a sake yeast with a mutated ARO4 gene which causes both resistance to ortho-fluoro-DL-phenylalanine and increased production of beta-phenethyl alcohol, J Fermentation Bioengin, 1992, **73**, 366

Fukuda K, Breeding of yeasts for producing large amounts of rose-like flavor components, Fragrance J, 1992, **20**, 35

Fukuda K, Watanabe M, Asano K, Ueda H, Ohta S, Breeding of brewing yeast producing a large amount of β-phenylethyl alcohol and β-phenylethyl acetate, Agric Biol Chem, 1990, **54** 269

Gold RS, Meagher MM, Hutkins R, Conway T, Ethanol tolerance and carbohydrate metabolism in *Lactobacilli*, J Ind Microbiol, 1992, **10**, 45

Götz F, Applied genetics in the gram positive bacterium *Staphylococcus carnosus*, Food Biotechnol, 1990, **4**, 505

Hadfield C, Raina KK, Shashimenon K, Mount RC, The expression and performance of cloned genes in yeasts, Mycol Res, 1993, **97**, 897

Hahm YT, Batt CA, Expression and secretion of thaumatin from *Aspergillus oryzae*, Agric Biol Chem, 1990, **54**, 2513

Hardjito L, Greenfield PF, Lee PL, Recombinant protein production via fed-batch culture of the yeast *Saccharomyces cerevisiae*, Enzyme Microbial Technol, 1993, **15**, 120

Hirata D, Aoki S, Watanabe K, Tsukioka M, Suzuki T, Stable overproduction of isoamyl alcohol by *Saccharomyces cerevisiae* with chromosome-integrated multicopy LEU4 genes, Biosci Biotech Biochem, 1992, **56**, 1682

Hoh YK, Tan TK, Yeoh HH, Protoplast fusion of beta-glucosidase-producing *Aspergillus niger* strains, Appl Biochem Biotechnol, 1992, **37**, 81

Ikeda M, Ozaki A, Katsumata R, Phenylalanine production by metabolically engineered *Corynebacterium glutamicum* with the phea gene of *Escherichia coli*, Appl Microbiol Biotechnol, 1993, **39**, 318

Inoue T, Sunagawa M, Mori A, Inai C, Fukuda M, Takagi M, Yano K, Cloning and sequencing of the gene encoding the 72KD dehydrogenase submit of alcohol dehydrogenase from *Acetobacter aceti*, J Bacteriol, 1989, **171**, 3115

Inoue Y, Kobayashi S, Yoshikawa K, Tran LT, Kimura A, Lipid hydroperoxide-resistance gene in *Saccharomyces cerevisiae* - utilization as a selectable marker gene for yeast transformation, Biotechnol Appl Biochem, 1993, **17**, 305

Ishibashi Y, Genetic studies into musty odor production by actinomycetes, Water Sci Technol, 1992, **25**, 171

Israelsen H, Hansen EB, Insertion of transposon Tn917 derivatives into the *Lactococcus lactis* subsp. *lactis* chromosome, Appl Environm Microbiol, 1993, **59**, 21

Jahns A, Schäfer A, Geis A, Teuber M, Identification, cloning and sequencing of the replication region of *Lactococcus lactis* ssp. *lactis* biovar. *diacetylactis* Bu2 citrate plasmid pSL2, FEMS Microbiol Let, 1991, **80**, 253

Javelot C, Girard P, Colonna-Ceccaldi B, Vladescu B, Introduction of terpene-producing ability in a wine strain of *Saccharomyces cerevisiae*, J Biotechnology, 1991, **21**, 239

Kawasumi T, Kiuchi N, Futatsugi Y, Ohba K, Yanagi So, High yield preparation of *Lentinus edodes* ('Shiitake') protoplasts with regeneration capacity and mating type stability, Agric Biol Chem, 1989, **51**, 1649

Kida K, Kume K, Morimura S, Sonoda Y, Repeated-batch fermentation process using a thermotolerant flocculating yeast constructed by protoplast fusion, J Fermentation Bioengin, 1992, **74**, 169

Klaenhammer TR, Development of bacteriophage resistant strains of lactic acid bacteria, Biochem Soc Trans, 1991, **19**, 675

Kok J, Genetics of proteolytic enzymes of *Lactococci* an their role in cheese flavor development, J Dairy Sci, 1993, **76**, 2056

Köttig H, Rottner G, Beck KF, Schweizer M, Schweizer E, The pentafunctional FAS 1 genes of *S. cerevisiae* and *Y. lipolytica* are co-linear and considerably longer than previously estimated, Mol Gen Genetics, 1991, **226**, 310

Kronlof J, Linko M, Production of beer using immobilized yeast encoding alpha-acetolactate decarboxylase, J Inst Brewing, 1992, **98**, 479

Kugimiya Q, Otani Y, Kohno M, Hashimoto Y, Cloning and sequence analysis of a DNA encoding *Rhizopus niveus* lipase, Agric Biol Chem, 1992, **56**, 716

Kumagai MH, Sverlow GG, Dellacioppa G, Grill LK, Conversion of starch to ethanol in a recombinant *Saccharomyces cerevisiae* strain expressing rice alpha-amylase from a novel *Pichia pastoris* alcohol oxidase promoter, Bio-Technology, 1993, **11**, 606

Kuriyama M, Morita S, Asakawa N, Nakatsu M, Kitano K, Stabilization of a recombinant plasmid in yeast, J Fermentation Bioeng, 1992, **74**, 139

Kusumegi K, Baba M, Nakahara T, Ogawa K, Breeding of soy sauce yeast by protoplast fusion technique. 1. Construction of protoplast fusants between soy sauce yeast and sake or wine yeast, Nippon Jozo Kyokaishi, 1992, **87**, 645

Law J, Fitzgerald GF, Uniacke-Lowe T, Daly C, Fox PF, The contribution of lactococcal starter proteinases to proteolysis in cheddar cheese, J Dairy Sci, 1993, **76**, 2455

Lonza AG, EP 477828 A2 920401, 1992

Mitter W, Kessler H, Biendl M, Attempts to obtain a reproducible hop aroma in beer, Brauwelt, 1993, **133**, 979 and 986

Mollet B, Constable A, Delley M, Knol J, Marciset O, Pridmore D, Molecular genetics in *Streptococcus thermophilus* from transformation to gene expression, Lait, 1993, **73**, 175

Nakamura S, Tagawa Y, Miyamoto T, Kataoka K, Plasmid profiles of lactose-, citrate- and proteinase-negative mutants of *Lactococcus lactis* subsp. *lactis* biovar *diacetylactis* strain N-7, Milk Sci Int, 1992, **47**, 358

Nowakowski CM, Bhowmilk TK, Steele JL, Cloning of peptidase genes from *Lactobacillus helveticus* CNRZ-32, Appl Microbiol Biotechnol, 1993, **39**, 204

O'Connell MJ, Kelly JM, Physical characterization of the aldehyde dehydrogenase encoding gene of *Aspergillus niger*, Gene, 1989, **84**, 173

Pan TM, Kuo CC, Production of flavor by a mutant of yeast, J Food Proc Pres, 1993, **17**, 213

Park S, Ryu DDY, Analysis of plasmid stability by the difference in colony size, J Fermentation Bioengin, 1992, **74**, 185

Patkar A, Seo JH, Fermentation kinetics of recombinant yeast in batch and fed-batch cultures, Biotechnol Bioengin, 1992, **40**, 103

Perez-Gonzalez JA, Gonzalez R, Querol A, Sendra J, Ramon D, Construction of a recombinant wine yeast strain expressing beta-(1,4)-endoglucanase and its use in microvinification processes, Appl Environ Microbiol, 1993, **59**, 2801

Pernod-Ricard, WO 9216611 A1 921001, 1992

Puta F, Wambutt R, Construction of a new *Escherichia coli Saccharomyces cerevisiae* shuttle plasmid cloning vector allowing positive selection for cloned fragments, Folia Microbiol, 1992, **37**, 193

Quest International B.V., Cloning and expression of DNA encoding a ripening form of a polypeptide having sulfhydryl oxidase activity, EP 565172 A1 931013, 1993

Ramakrishnan S, Hartley BS, Fermentation of lactose by yeast cells secreting recombinant fungal lactase, Appl Environ Microbiol, 1993, **59**, 4230

Redfield RJ, Genes for breakfast - the have-your-cake-and-eat-it-too of bacterial transformation, J Heredity, 1993, **84**, 40

Requena T, McKay LL, Plasmid profiles and relationship to lactose utilization proteinase activity in a lactococcal strain isolated from semi-hard natural cheese, Milk Sci Int, 1993, **48**, 264

Romanos MA, Scorer CA, Clare JJ, Foreign gene expression in yeast - a review, Yeast, 1992, **8**, 423

Romero DA, Klaenhammer TR, Transposable elements in *Lactococci* - a review, J Dairy Sci, 1993, **76**, 1

Saito K, Watanabe S, Taguchi T, Takahashi H, Nakata T, Iwano K, Ishikawa T, Studies on brewing of a ginjo-sake. 1. Isolation of sake-yeast mutants having high ester-producing ability by using koji extract medium supplemented with alcohol dehydrated koji and properties of the mutants, Nippon Jozo Kyokaishi, 1992, **87**, 915

Schnurr G, Schmidt A, Sandmann G, Mapping of a carotenogenic gene cluster from *Erwinia herbicola* and functional identification of six genes, FEMS Microbiol Lett, 1991, **78**, 157

Sebastiao MJ, Cabral JMS, Airesbarros MR, Synthesis of fatty acid esters by a recombinant cutinase in reversed micelles, Biotechnol Bioengin, 1993, **42**, 326

Sesma F, Gardiol D, de Ruiz Holgado AP, de Mendoza D, Cloning of the citrate permease gene of *Lactococcus lactis* subsp. *lactis* biovar *diacetylactis* and expression in *Escherichia coli*, Appl Environm Microbiol, 1990, **56**, 2099

Shibuya I, Tamura G, Goto E, Ishikawa T, Hara S, Characteristics of koji prepared from the transformant of *Aspergillus oryzae* with the glucoamylase gene of *Aspergillus shirousamii*, and its utilization for sake-brewing, J Fermentation Bioeng, 1992, **73**, 415

Somkuti GA, Solaiman DKY, Steinberg DH, Expression of *Streptomyces* sp cholesterol oxidase in *Lactobacillus casei*, Appl Microbiol Biotechnol, 1992, **37**, 330

Teuber M, Genetic engineering techniques in food microbiology and enzymology, Food Rev Int, 1993, **9**, 389

Unilever NV, EP 487159 A1 920527, 1992

van de Guchte M, Kok J, Venema G, Gene expression in *Lactococcus lactis*, FEMS Microbiol Rev, 1992, **88**, 73

Ward M, Wilson LJ, Kodama KH, Use of *Aspergillus* overproducing mutants, cured for integrated plasmid, to overproduce heterologous proteins, Appl Microbiol Biotechnol, 1993, **39**, 738

Watanabe M, Fukuda K, Asano K, Ohta S, Mutants of bakers yeasts producing a large amount of isobutyl alcohol or isoamyl alcohol, flavour components of bread, Appl Microbiol Biotechnol, 1990, **34**, 154

Watanabe M, Tanaka N, Mishima H, Takemu S, Isolation of sake yeast mutants resistant to isoamyl monofluoroacetate to improve isoamyl acetate productivity, J Fermentation Bioeng, 1993, **76**, 229

Wells JM, Wilson PW, Lepage RWF, Improved cloning vectors and transformation procedure for *Lactococcus lactis*, J Appl Bacteriol, 1993, **74**, 629

Wightman JD, Xu X, Yorgey BM, Watson BT, McDaniel MR, Micheals NJ, Bakalinsky AT, Evaluation of genetically modified wine strains of *Saccharomyces cerevisiae*, Amer J Enol Vitic, 1992, **43**, 283

Wood BE, Ingram LO, Ethanol production from cellobiose, amorphous cellulose, and crystalline cellulose by recombinant *Klebsiella oxytoca* containing chromosomally integrated *Zymomonas mobilis* genes for ethanol production and plasmids expressing thermostable cellulase genes from *Clostridium thermocellum*, Appl Environ Microbiol, 1992, **58**, 2103

Yamaguchi S, Mase T, Takeuchi K, Cloning and structure of the mono- and diacylglycerol lipase-encoding gene from *P. camembertii* U-150, Gene, 1991, **103**, 61

Yoshida K, Inahashi M, Noro F, Nakamura K, Nojiro K, Breeding of a sake yeast having high productivities of flavor esters and low acid productivity, Nippon Jozo Kyokaishi, 1993, **88**, 565

Yoshikawa S, Kondo K, Kuwabara H, Arika Y, Oguri I, Okazaki M, Engineering of salt-tolerant yeasts by mutation. Part 3. Studies on mutation system of salt-tolerant yeasts producing beta-phenylethyl alcohol, miso flavor components, Nagano-ken Shokuhin Kogyo Shikenjo Kenkyu Hokoku, 1992, **20**, 51

Chapter 9

Addo K, Burton D, Stuart MR, Burton HR, Hildebrand DF, Soybean flour lipoxygenase isozyme mutant effects on bread dough volatiles, J Food Sci, 1993, **58**, 583

Agrawal R, Patwardhan MV, Gurudutt KN, Formation of flavouring constituents in callus cultures of lime, *Citrus aurantifolia* S, Biotech Appl Biochem, 1991, **14**, 265

Ahn SN, Bollich CN, Tanksley SD, RFLP Tagging of a gene for aroma in rice, Theor Appl Genetics, 1992, **84**, 825

Akakabe Y, Naoshima Y, The mechanistic pathway of the biotransformation of acetophenone by immobilized cell cultures of *Gardenia*, Phytochemistry, 1993, **32**, 1189

Almosnino AM, Belin JM, Apple pomace: an enzyme system for producing aroma compounds from polyunsaturated fatty acids, Biotechnol Let, 1991, **13**, 893

Andreev VY, Vartapetian BB, Induction of alcoholic and lactic fermentation in the early stages of anaerobic incubation of higher plants, Phytochemistry, 1992, **31**, 1859

Arigoni D, Cane DE, Shim JH, Croteau R, Wagschal K, Monoterpene cyclization mechanisms and the use of natural abundance deuterium NMR - short cut or primrose path, Phytochemistry, 1993, **32**, 623

Asada Y, Saito H, Yoshikawa T, Sakamoto K, Furuya T, Studies on plant tissue culture 88. biotransformation of 18 beta-glycyrrhetinic acid by ginseng hairy root culture, Phytochemistry, 1993, **34**, 1049

Askar A, Bielig HJ, Geschmacksverbesserung von Lebensmitteln, Alimenta, 1976, **15**, 155

Bajaj YPS, Reghunath BR, Gopalakrishnan PK, *Elettaria cardamomum maton* (cardamom): aromatic compounds, *vitro* culture studies, and clonal propagation, Biotechnol Agric For, 21 (Medicinal and Aromatic Plants IV), 1993, 132

Beimen A, Witte L, Barz W, Growth characteristics and elicitor-induced reactions of photosynthetically active and heterotrophic cell suspension cultures of *Lycopersicon peruvianum* (Mill), Botanica Acta, 1992, **105**, 152

Berger RG, Akkan Z, Drawert F, Catabolism of geraniol by suspension cultured *C. limon* cv. *Ponderosa*, Biochem Biophys Acta, 1991, **1055**, 234

Berger RG, Biotechnical production of flavours - current status, In: Understanding natural flavors, Piggott JR, Paterson A (eds), Chapman & Hall London 1994, 178

Berger RG, Dettweiler GR, Krempler GMR, Drawert F, Precursor atmosphere technology - Efficient aroma enrichment in fruits, In: Flavor Precursors, Teranishi R, Takeoka G, Güntert M (eds), ACS symp ser 490, ACS Wash DC 1992, 59

Berger RG, Drawert F, Kollmannsberger H, PA-Lagerung zur Kompensation von Aromaverlusten bei der Gefriertrocknung von Bananenscheiben, Z Lebensm Unters Forsch, 1986, **183**, 169

Brackmann A, Streif J, Bangerth F, Relationship between a reduced aroma production and lipid metabolism of apples after long-term controlled-atmosphere storage, J Amer Soc Horticult Sci, 1993, **118**, 243

Brodelius P, Pedersen H, Increasing secondary metabolite production in plant-cell culture by redirecting transport, Trends Biotechnol, 1993, **11**, 30

Bruchmann EE, Kolb E, Wirkung von Inhibitoren auf das 'Aromaenzymsystem' der Himbeerkelchzapfen, Lebensm-Wiss Technol, 1973, **6**, 107

Buitelaar RM, Tramper J, Strategies to improve the production of secondary metabolites with plant cell cultures - a literature review, J Biotechnology, 1992, **23**, 111

Calvo MD, Sanchez-Gras MC, Accumulation of monoterpenes in shoot-proliferation cultures of *Lavandula latifolia* Med, Plant Science, 1993, **91**, 207

Collin HA, Britton G, *Allium cepa* L. (onion): *vitro* culture and the production of flavor, Biotechnol Agric For, 24 (Medicinal and Aromatic Plants V) **23**, 1993, Comai L, Impact of plant genetic engineering on foods and nutrition, Ann Rev Nutrit, 1993, **13**, 191

Crouzet J, La régénération enzymatique des arômes, Bios, 1977, **8**, 29

De Billy F, Paupardin C, Sur l'évolution des huiles essentielles dans les tissus de péricarpe de citron (*Citrus limonia* Obseck) cultives *vitro* CR Acad Sci Paris Ser D, 1971, **273**, 1693

Del Rio JA, Ortuno A, Marin FR, Puig DG, Sabater F, Bioproduction of neohesperidin and naringin in callus cultures of *Citrus aurantium*, Plant Cell Rep, 1992, **11**, 592

Doran PM, Design of reactors for plant cells and organs, In: Biochemical Engineering Biotechnology, Fiechter A (ed), Springer Berlin 1993, **48**, 115

Fauconnier ML, Jaziri M, Marlier M, Roggemans J, Wathelet JP, Lognay G, Severin M, Homes J, Shimomura K, Essential oil production by *Anthemis nobilis* L tissue culture, J Plant Physiology, 1993, **141**, 759

Flesch V, Jacques M, Cosson L, Teng BP, Petiard V, Barz JP, Relative importance of growth and light level on terpene content of *Ginkgo biloba*, Phytochemistry, 1992, **31**, 1941

Fujita Y, Tabata M, Plant tissue and cell culture, In: Plant Biology, Green CE, Somers DA, Hacket WP, Biesboer DD (eds), Liss New York 1987, 169

Funk C, Croteau R, Induction and characterization of a cytochrome-p-450-dependent camphor hydroxylase in tissue cultures of common sage (*Salvia officinalis*), Plant Physiol, 1993, **101**, 1231

Gbolade AA, Lockwood GB, Growth and production of volatile substances by *Melissa officinalis* and *Petroselinum crispum* cultures, Fitoterapia, 1991, **62**, 237

Gbolade AA, Lockwood GB, Some factors affecting productivity of *Allium cepa* L callus cultures, J Ess Oil Res, 1992, **4**, 381

German JB, Berger RG, Drawert F, Generation of fresh fish flavor: Rainbow trout (*Salmo gairdneri*) gill homogenate as a model system, Chem Mikrobiol Technol Lebensm, 1991, **13**, 19

German JB, Zhang H, Berger RG, Role of lipoxigenases in lipid oxidation in foods, In: Lipid Oxidation in Food, Angelo AJSt (ed), ACS symp ser 500, ACS Wash DC 1992, 74

Guadagni DG, Bomben JL, Hudson JS, Factors influencing the development of aroma in apple peels, J Sci Fd Agric, 1971, **22**, 110

Hansen K, Poll L, Olsen CE, Lewis MJ, The influence of oxygen concentration in storage atmospheres on the post-storage volatile ester production of *Jonagold* apples, Food Sci Technol - Lebensm Wiss Technol, 1992, **25**, 457

Hatanaka A, The biogeneration of green odour by green leaves, Phytochemistry, 1993, **34**, 1201

Hayashi H, Yamada K, Fukui H, Tabata M, Metabolism of exogenous 18beta-glycyrrhetinic acid in cultured cells of *Glycyrrhiza glabra*, Phytochemistry, 1992, **31**, 2729

Hewitt EJ, MacKay DAM, Königsbacher K, Hasselstrom T, The role of enzymes in food flavors, Food Technol, 1956, **10**, 487

Hildebrand DF, Altering fatty acid metabolism in plants, Food Technology, 1992, **46**, 71

Hook I, Sheridan H, Wilson G, Volatile metabolites from suspension cultures of *Taraxacum officinale*, Phytochemistry, 1991, **30**, 3977

Ilahi I, Jabeen M, Tissue culture studies for micropropagation and extraction of essential oils from *Zingiber officinale* Rosc, Pakistan J Bot, 1992, **24**, 54

Jain SM, Pehu E, The prospects of tissue culture and genetic engineering for strawberry improvement - review article, Acta Agricult Scandinavica Sect B, 1992, **42**, 133

Joersbo M, Enevoldsen K, Andersen JM, Oil glands on black currant callus, J Plant Physiology, 1992, **140**, 96

Josephson DB & Lindsay RC, Enzymic generation of volatile aroma compounds from fresh fish, In: Biogeneration of Aromas, Parliment TH, Croteau R (eds), ACS symp ser 317, ACS Wash DC 1986, 201

Kamel S, Brazier M, Desmet G, Fliniaux MA, Jacquin-Dubreuil A, Glucosylation of butyric acid by cell suspension culture of *Nicotiana plumbaginifolia*, Phytochemistry, 1992, **31**, 1581

Kane EJ, Wilson AJ, Chourey PS, Mitochondrial genome variability in sorghum cell culture protoclones, Theor Appl Genetics, 1992, **83**, 6

Kawabe S, Fujiwara H, Murakami K, Hosomi K, Volatile constituents of *Mentha arvensis* cultures, Biosci Biotech Biochem, 1993, **57**, 657

Kennedy AI, Deans SG, Svoboda KP, Gray AI, Waterman PG, Volatile oils from normal and transformed root of *Artemisia absinthium*, Phytochemistry, 1993, **32**, 1449

Kennedy RA, Rumpho ME, Fox TC, Anaerobic metabolism in plants, Plant Physiology, 1992, **100**, 1

Kesselring K, Pflanzenzellkulturen zur Auffindung neuer, therapeutisch relevanter Naturstoffe und deren Gewinnung durch Fermentationsprozesse, In: Pflanzliche Zellkulturen, BMFT-Statusseminar, Bundesministerium für Forschung und Technologie (BMFT)(ed), Bonn 1985, 111

Kitamura K, Ishimoto M, Kikuchi A, Kaizuma N, A new soybean variety 'Yumeyutaka' lacking both L-2 and L-3 of the lipoxygenase isozymes, Jap J Breeding, 1992, **42**, 905

Knorr D, Beaumont MD, Caster CS, Dörnenburg H, Gross B, Pandya Y, Romagnoli LG, Plant tissue culture for the production of naturally derived food ingredients, Food Technol 1990, **44**, 71

Knorr D, Caster C, Doerneburg H, Dorn R, Graef S, Havkin-Frenkel D, Podstolski A, Werrman U, Biosynthesis and yield improvement of food ingredients from plant cell and tissue cultures, Food Technol 1993, **47**, 57

Kometani T, Tanimoto H, Nishimura T, Okada S, Glucosylation of vanillin by cultured plant cells, Biosci Biotech Biochem, 1993, **57**, 1290

Kurata H, Furusaki S, Immobilized *Coffea arabica* cell culture using a bubble-column reactor with controlled light intensity, Biotechol Bioengin, 1993, **42**, 494

Lipsky AK, Problems of optimisation of plant cell culture processes, J Biotechnology, 1992, **26**, 83

Lizotte PA, Shaw PE, Flavor volatiles in *Valencia* orange and their quantitative changes caused by vacuum infiltration of butanal, Food Sci Technol - Lebensm Wiss Technol, 1992, **25**, 80

Loreto F, Sharkey TD, Isoprene emission by plants is affected by transmissible wound signals, Plant Cell and Environment, 1993, **16**, 563

Mars GB , WO 9119800 A1 911226, 1991

Mattheis JP, Roberts RG, Fumigation of sweet cherry (*Prunus avium* Bing) fruit with low molecular weight aldehydes for postharvest decay control, Plant Disease, 1993, **77**, 810

McDowall RM, Clark BM, Wright GJ, Northcote TG, Trans-2-cis-6-nonadienal - The cause of cucumber odor in osmerid and retropinnid smelts, Trans Amer Fisheries Soc, 1993, **122**, 144

McKie JH, Jaouhari R, Douglas KT, Goffner D, Feuillet C, Grima-Pettenati J, Boudet AM, Baltas M, Gorrichon L, A molecular model for cinnamyl alcohol dehydrogenase, a plant aromatic alcohol dehydrogenase involved in lignification, Biochim Biophys Acta, 1993, **1202**, 61

Mulder-Krieger TH, Verpoorte R, Baerheim Svendsen A, Scheffer J, Production of essential oils and flavours in plant cell and tissue cultures, A Review, Plant Cell, Tissue and Organ Culture, 1988, **13**, 85

Muller BL, Gautier AE, Green note production: a challenge for biotechnology, In: Trends in flavour research, Maarse H, van der Heij DG (eds), Elsevier Amsterdam 1994, 475

Nabeta K, Kawachi J, Sakurai M, Volatile production in cultured cells and plantlet regeneration from protoplast of patchouli, *Pogostemon cablin*, Dev Food Sci, 1993, **32**, 577

Nabeta K, Kawakita K, Yada Y, Okuyama H, Biosynthesis of sesquiterpenes from deuterated mevalonates in *Perilla* callus, Biosci Biotech Biochem, 1993, **57**, 792

Nanos GD, Romani RJ, Kader AA, Metabolic and other responses of *Bartlett* pear fruit and suspension cultured *Passa Crassanae* pear fruit cells held in 0.25% O_2, J Amer Soc Horticult Sci 1992, 117, 934

Noma Y, Akehi E, Miki N, Asakawa Y, Biotransformation of terpene aldehydes, aromatic aldehydes and related compounds by *Dunaliella tertiolecta*, Phytochemistry, 1992, **31**, 515

Noma Y, Asakawa Y, Enantio- and diastereoselectivity in the biotransformation of carveols by *Euglena gracilis* Z, Phytochemistry, 1992, **31**, 2009

Ohsumi C, Hayashi T, Sano K, Formation of alliin in the culture tissues of *Allium sativum* - Oxidation of S-Allyl-L-Cysteine, Phytochemistry, 1993, **33**, 107

Ono K, Toyota M, Asakawa Y, Constituents from cell suspension cultures of selected liverworts, Phytochemistry, 1992, **31**, 1249

Orihara Y, Furuya T, Studies on plant tissue culture. 58. Biotransformation of (-)-borneol by cultured cells of *Eucalyptus perriniana*, Phytochemistry, 1993, **34**, 1045

Ozcan S, Firek S, Draper J, Selectable marker genes engineered for specific expression in target cells for plant transformation, Bio-Technology, 1993, **11**, 218

Para G, Baratti J, Irones and precursors synthesized by *Iris sibirica* tissue culture, Biosci Biotech Biochem, 1992, **56**, 1132

Paterson A, Piggott JR, Jiang J, Approaches to mapping loci that influence flavor quality in raspberries, In: Food Flavor and Safety, ACS symp ser **528**, ACS Wash DC 1993, 109

Pawlowicz P, Wawrzenczyk C, Siewinski A, Biotransformations. 31. Uncommon type of hydroxylation of 3-alkenylsubstituted derivatives of citronellol and citronellic acid by *Spirodela punctata*, Phytochemistry, 1992, **31**, 2355

Penarrubia L, Kim R, Giovannoni J, Kim SH, Fischer RL, Production of the sweet protein monellin in transgenic plants, Bio-Technology, 1992, **10**, 561

Pereira M, Moreira MA, Rezende ST, Sediyama CS, Effects of linoleic free acid, methyl linoleate and trilinolein on the production of carbonyl compounds in soybean seed homogenates, Arquivos Biologia Tecnologia, 1992, **35**, 403

Perez AG, Rios JJ, Sanz C, Olias JM, Aroma components and free amino acids in strawberry variety *Chandler* during ripening, J Agric Food Chem, 1992, **40**, 2232

Perez AG, Sanz C, Olias JM, Partial purification and some properties of alcohol acyltransferase from strawberry fruits, J Agric Food Chem, 1993, **41**, 1462

Pernod-Ricard, FR Pat Appl, 8912901, 1991

Prince C, Shuler ML, Manufacture of sulfur-containing flavor components and precursors with cultured root tissue of *Allium*, US, 5244794A 930914, 1993

Procter & Gamble, US 3,224,886, 1965

Reil G PhD Thesis TU Munich, 1993

Reil G, Akkan Z, Berger RG, *Coleonéma album*: *In vitro* culture and the production of essential oils, In: Medicinal and Aromatic Plants, Bajaj YPS (ed), Springer Berlin, in preparation

Sakui N, Kuroyanagi M, Ishitobi Y, Sato M, Ueno A, Biotransformation of sesquiterpenes by cultured cells of *Curcuma zedoaria*, Phytochemistry, 1992, **31**, 143

Schreier P 1984, Chromatographic studies of biogenesis of plant volatiles, Chromatogr Methods, Hüthig Heidelberg

Schultze W, Hose S, Abou-Mandour A, Czygan FC, *Melissa officinalis* L (lemon balm): in vitro culture and the production and analysis of volatile compounds, Biotechnol Agric For, 24 (Medicinal and Aromatic Plants V), 1993, 242

Scragg AH, Arias-Castro C, Bioreactors for industrial production of flavours: Use of plant cells, In: Bioformation of Flavours, Patterson RLS, Charlwood BV, MacLeod G, Williams AA (eds), Royal Soc Chem Cambridge 1992, 131

Shamaila M, Powrie WD, Skura BJ, Analysis of volatile compounds from strawberry fruit stored under modified atmosphere packaging (MAP), J Food Sci, 1992, **57**, 1173

Shaw PE, Moshonas MG, Nisperos Carriedo MO, Carter RD, Controlled atmosphere treatment of freshly harvested oranges at elevated temperature to increase volatile flavor components, J Agric Food Chem, 1992b, **40**, 1041

Shaw PE, Moshonas MG, Nisperos Carriedo MO, Controlled atmosphere storage effects on the composition off volatile components in dancy mandarin and mandarin hybrid fruit, Food Sci Technol - Lebensm Wiss Technol, 1992a, **25**, 346 and 236

Spencer A, Hamill JD, Rhodes MJ, *In vitro* biosynthesis of monoterpenes by *Agrobacterium* transformed shoot cultures of 2 *Mentha* species, Phytochemistry, 1993, **32**, 911

Stöckigt J, Schübel N, Naturstoffe aus pflanzlichen Zellkulturen, Dtsch Apotheker Ztg, 1989, **129**, 1187

Suga T, Tang YX, Selective hydrogenation of the C-C double bond of alpha, beta-unsaturated carbonyl compounds by the immobilized cells of *Nicotiana tabacum*, J Nat Prod, 1993, **56**, 1406

Tabata M, Jujita YM, Production of shikonin by plant cell cultures, In: Biotechnology in Plant Science, Zatlin M, Day P, Hollaender A, (eds), Academic Press Orlando FLA 1985, 207

Tamura H, Takebayashi T, Sugisawa H, Thymus vulgaris L (thyme): in vitro culture and the production of secondary metabolites, Biotechnol Agric For, 21 (Medicinal and Aromatic Plants IV), 1993, 413

Tani M, Takeda K, Yazaki K, Tabata M, Effects of oligogalacturonides on biosynthesis of shikonin in *Lithospermum* cell cultures, Phytochemistry, 1993, **34**, 1285

Toivonen L, Utilization of hairy root cultures for production of secondary metabolites - topical paper, Biotechnol Prog, 1993, **9**, 12

Ueda Y, Bai JH, Yoshioka H, Effects of polyethylene packaging on flavor retention and volatile production of *Starking Delicious* apple, J Jap Soc Horticult Sci, 1993, **62**, 207

Werrmann U, Plant cell cultures as model system for biosynthetic pathways? Conversion of monoterpenes in *Mentha* cell suspension cultures, In: Progress in Flavour Precursor Studies, Schreier P, Winterhalter P (eds), Allured Carol Stream IL 1992, 195

Westcott RJ, Cheetham PSJ, Barraclough AJ, Use of organized viable vanilla plant aerial roots for the production of natural vanillin, Phytochemistry, 1994, **35**, 135

Weurman C, Gas liquid chromatographic studies on the enzymatic formation of volatile compounds in raspberries, Food Technol, 1961, **15**, 531

Yoshikawa T, Asada Y, Furuya T, Studies on plant tissue culture 84. Continuous production of glycosides by a bioreactor using *Ginseng* hairy root culture, Appl Microbiol Biotechnol, 1993, **39**, 460

Zhang CH, Hirano T, Suzuki T, Shirai T, Morishita A, Studies on the odor of fish. III. Volatile compounds and their generation in smelt, Nippon Suisan Gakkaishi, 1992, **58**, 773

Chapter 10

Aarts RJ, Possibilities to improve the process in food- and biotechnology, Food Biotechnol, 1990, **4**, 301

Abraham B, Krings U, Berger RG, Dynamic extraction, an efficient screening procedure for aroma producing basidiomycetes, Chem Mikrobiol Technol Lebensm, 1993, **15**, 178

Abraham TE, Surender GD, Horizontal rotary bioreactor - effect of rotation and bed characteristics on ethanol fermentation with immobilized yeast cells, J Fermentation Bioeng, 1993, **75**, 322

Antier P, Minjares A, Roussos S, Viniegra-Gonzalez G, New approach for selecting pectinase producing mutants of *Aspergillus niger* well adapted to solid state fermentation, Biotechnol Adv, 1993, **11**, 429

Araki K, Minami T, Sueki M, Kimura T, Optimum operational conditions of a 2-stage-extraction butanol extractive fermentation process and continuous fermentation by butanol-isopropanol producing microorganisms immobilized by Ca-alginate, Seibutsu - Kogaku Kaishi - J Soc Fermentation Bioengi, 1993, **71**, 9 and 389

Auria R, Morales M, Villegas E, Revah S, Influence of mold growth on the pressure drop in aerated solid state fermentors, Biotechnol Bioeng, 1993, **41**, 1007

Azuma S, Tsunekawa H, Okabe M, Okamoto R, Aiba S, Hyper-production of L-tryptophan via fermentation with crystallization, Appl Microbiol Biotechnol, 1993, **39**, 471

Bare G, Gerard J, Jacques P, Delaunois V, Thonart P, Bioconversion of vanillin into vanillic acid by *Pseudomonas fluorescens* strain BTP9 - cell reactors and mutants study, Appl Biochem Biotechnol, 1992, **34**, 499

Beaumelle D, Marin M, Gibert, Pervaporation of aroma compounds in water ethanol mixtures - experimental analysis of mass transfer, J Food Eng, 1992, **16**, 293

Bejar P, Casas C, Godia F, Sola C, The influence of physical properties on the operation of a 3-phase fluidized-bed fermentor with yeast cells immobilized in Ca-alginate, Appl Biochem Biotechnol, 1992, **34**, 467

Bengtsson E, Tragardh G, Hallstrom B, Concentration polarization during the enrichment of aroma compounds from a water solution by pervaporation, J Food Eng, 1993, **19**, 399

Berger RG, Biotechnical production of flavours, In: Understanding Natural Flavours, Piggott JR, Paterson A (eds), Chapman & Hall London 1994, 178

Berger RG, Drawert F, Schlebusch JP, Fermentation technologies for flavour producing cell cultures, In: Engineering & Food, Proc Intern Congr Köln, Spieß WEL, Schubert H (eds), Elsevier London New York 1990, **3**, 455

Boyaval P, Seta J, Gavach C, Concentrated propionic acid production by electrodialysis, Enzyme Microbial Technol, 1993, **15**, 683

Buitelaar RM, Leenen EJTM, Geurtsen G, de Groot E, Effects of the addition of XAD-7 and of elicitor treatment on growth, thiophene production, and excretion by hairy roots of *Tagetes patula*, Enzyme Microbial Technol, 1993, **15**, 670

Byers JP, Shah MB, Fournier RL, Varanasi S, Generation of a pH gradient in an immobilized enzyme system, Biotechnol Bioeng, 1993, **42**, 410

Cai YJ, Buswell JA, Chang ST, Effect of lignin-derived phenolic monomers on the growth of the edible mushrooms *Lentinus edodes*, *Pleurotus sajor-caju* and *Volvariella volvacea*, World J Microbiol Biotechnol, 1993, **9**, 503

Casillas JL, Addoyobo F, Kenney CN, Aracil J, Martinez M, The use of modified divinylbenzene polystyrene resins in the separation of fermentation products - a case study utilizing amino acids and a dipeptide, J Chem Technol Biotechnol, 1992, **55**, 163

Chang HN, Lee WG, Kim BS, Cell retention culture with an internal filter module - continuous ethanol fermentation, Biotechnol Bioeng, 1993, **41**, 677

Christen P, Auria R, Vega C, Villegas E, Revah S, Growth of *Candida utilis* in solid state fermentation, Biotechnol Adv, 1993, **11**, 549

Colomban A, Roger L, Boyaval P, Production of propionic acid from whey permeate by sequential fermentation, ultrafiltration, and cell recycling, Biotechnol Bioeng, 1993, **42**, 1091

Creuly C, Larroche C, Gros JB, Bioconversion of fatty acids into methyl ketones by spores of *Penicillium roqueforti* in a water organic solvent, 2-phase system, Enzyme Microbial Technol, 1992, **14**, 669

Cubarsi R, Villaverde A, Conditions for a continuous production of transient microbial products in a 2-stage fermentation system, Biotechnol Let, 1993, **15**, 827

Dale MC, Okos MR, Wankat PC, An immobilized cell reactor with simultaneous product separation. I. Reactor design and analysis, II. Experimental reactor performance, Biotechnol Bioeng, 1985, **27**, 932 and 943

Davison BH, Thompson JE, Continuous direct solvent extraction of butanol in a fermenting fluidized-bed bioreactor with immobilized *Clostridium acetobutylicum*, Appl Biochem Biotechnol, 1993, **39**, 415

Debacker L, Devleminck S, Willaert R, Baron G, Reaction and diffusion in a gel membrane reactor containing immobilized cells, Biotechnol Bioeng, 1992, **40**, 322

Debus O, Wanner O, Degradation of xylene by a biofilm growing on a gas-permeable membrane, Water Sci Technol, 1992, **26**, 607

Dos Santos LMF, Livingston AG, A novel bioreactor system for the destruction of volatile organic compounds, Appl Microbiol Biotechnol, 1993, **40**, 151

Drouin CM, Cooper DG, Biosurfactants and aqueous 2-phase fermentation, Biotechnol Bioeng, 1992, **40**, 86

Duff SJB, Murray WD, Development of a microbially catalyzed oxidation system, Dev Food Sci, 1992, **29**, 1

Ellis J, Korth W, Removal of geosmin and methylisoborneol from drinking water by adsorption on ultrastable zeolite-Y, Water Research, 1993, **27**, 535

Eyal AM, Bressler E, Industrial separation of carboxylic and amino acids by liquid membranes - applicability, process considerations, and potential advantages - mini-review, Biotechnol Bioeng, 1993, **41**, 287

Fischer U PhD Thesis U Hannover 1994

Font G, Mañes J, Moltó JC, Picó Y, Solid-phase extraction in multi-residue pesticide analysis of water, J Chromatography, 1993, **642**, 135

Freeman A, Woodley JM, Lilly MD, *In situ* product removal as a tool for bioprocessing, Bio-Technology, 1993, **11**, 1007

Garcia HS, Malcata FX, Hill CG, Amundson CH, Use of *Candida rugosa* lipase immobilized in a spiral wound membrane reactor for the hydrolysis of milkfat, Enzyme Microbial Technol, 1992, **14**, 535

Ghildyal NP, Ramakrishna M, Lonsane BK, Karanth NG, Gaseous concentration gradients in tray type solid state fermentors - effect on yields and productivities, Bioprocess Eng, 1992, **8**, 67

Gilson CD, Thomas A, A novel fluidised bed bioreactor for fermentation of glucose to ethanol using alginate immobilised yeast, Biotechnol Techn, 1993, **7**, 397

Gowthaman MK, Ghildyal NP, Rao KSMSR, Karanth NG, Interaction of transport resistances with biochemical reaction in packed bed solid state fermenters - the effect of gaseous concentration gradients, J Chem Technol Biotechnol, 1993b, **56**, 233

Gowthaman MK, Rao KSMSR, Ghildyal NP, Karanth NG, Gas concentration and temperature gradients in a packed bed solid-state fermentor, Biotechnol Adv, 1993a, **11**, 611

Grobben NG, Eggink G, Cuperus FP, Huizing HJ, Production of acetone, butanol and ethanol (ABE) from potato wastes - fermentation with integrated membrane extraction, Appl Microbiol Biotechnol, 1993, **39**, 494

Groot WJ, Kraayenbrink MR, Vanderlans RGJM, Luyben KCAM, Ethanol production in an integrated fermentation membrane system - process simulations and economics, Bioprocess Eng, 1993, **8**, 189

Groot WJ, Kraayenbrink MR, Waldram RH, Vanderlans RGJM, Luyben KCAM, Ethanol production in an integrated process of fermentation and ethanol recovery by pervaporation, Bioprocess Eng, 1992a, **8**, 99

Groot WJ, Sikkenk CM, Waldram RH, Vanderlans RGJM, Luyben KCAM, Kinetics of ethanol production by bakers yeast in an integrated process of fermentation and microfiltration, Bioprocess Eng, 1992, **8**, 39

Groot WJ, Vanderlans RGJM, Luyben KCAM, Technologies for butanol recovery integrated with fermentations, Process Biochem, 1992b, 27, 61

Gumbirasaid E, Greenfield, PF, Mitchel DA, Doelle HW, Operational parameters for packed beds in solid-state cultivation, Biotechnol Adv, 1993, **11**, 599

Halligudi SB, Khan MMT, Desai NV, Rangarajan R, Rao AV, Ultrafiltration membranes for the separation of catalyst and products in homogeneous catalysis systems, J Chem Technol Biotechnol, 1992, **55**, 313

He RQ, Xu J, Li CY, Zhao XA, The principle of parabolic feed for a fed-batch culture of baker's yeast, Appl Biochem and Biotechnol, 1993, **41**, 145

Hector ML, Murphy-Waldorf MF, Giertych TB, Hickey MJ, Haggard AA, Isolation and characterization of *Pseudomonas aeruginosa* mutants deficient in the utilization of the terpenoid citronellic acid, World J Microbiol Biotechnol, 1993, **9**, 562

Heinzle E, Present and potential applications of mass spectrometry for bioprocess research and control, J Biotechnology, 1992, **25**, 81

Ishizaki A, Takasaki S, Furuta Y, Cell-recycled fermentation of glutamate using a novel cross-flow filtration system with constant air supply, J Fermentation Bioeng, 1993, **76**, 316

Ito M, Yoshikawa S, Asami K, Hanai T, Dielectric monitoring of gas production in fermenting bread dough, Cereal Chem, 1992, **69**, 325

Jarzebski AB, Malinowski JJ, Goma G, Soucaille P, Analysis of continuous fermentation processes in aqueous 2-phase system, Bioprocess Eng 1992, **7**, 315

Jones TD, Havard JM, Daugulis AJ, Ethanol production from lactose by extractive fermentation, Biotechnol Let, 1993, **15**, 871

Kamal A, Reddy BSP, Enzymatic regeneration of aldehydes and ketones from hydrazones - oximes by baker's yeast in organic media, Biotechnol Let, 1992, **14**, 929

Kapfer GF, Berger RG, Drawert F, Improved production of 4-decanolide by biomass recirculation, Chem Mikrobiol Technol Lebensm, 1991, **13**, 1

Kapfer GF, Krüger H, Nitz S, Drawert F, Gewinnung von fermentativ hergestellten Aromen durch Adsorption und CO_2-Hochdruckextraktion, ZFL 1990, **41**, 51

Kerem Z, Hadar Y, Chemically defined solid-state fermentation of *Pleurotusostreatus*, Enzyme Microbial Technol, 1993, 15, 785

Kim YJ, Weigand WA, Experimental analysis of a product inhibited fermentation in an aqueous 2-phased system, Appl Biochem Biotechnol, 1992, **34**, 419

Kiyohara H, Hatta T, Ogawa Y, Kakuda T, Yokoyame H, Takizawa N, Isolation of *Pseudomonas pickettii* strains that degrade 2,4,6-trichlorophenol and the dechlorination of chlorophenols, Appl Environm Microbiol, 1992, **58**, 1276

Klingenberg A, Hanssen HP, Enhanced production of volatile flavor compounds from yeasts by adsorber techniques, Chem Biochem Eng Q, 1988, **2**, 222

Krahn U PhD Thesis TU Berlin 1986

Kramer CM, Kory MM, Bacteria that degrade para-chlorophenol isolated from a continuous culture system, Can J Microbiol, 1992, **38**, 34

Krings U, Berger RG, 1994, Porous polymers for fixed bed adsorption of aroma compounds in fermentation processes, Biotech Techn, 1995, **9**, 19

Krings U, Kelch M, Berger RG, Adsorbents for the recovery of aroma compounds in fermentation processes, J Chem Tech Biotechnol, 1993, **58**, 293

Kuhad RC, Singh A, Enhanced production of cellulases by *Penicillium citrinum* in solid state fermentation of cellulosic residue, World J Microbiol Biotechnol, 1993, **9**, 100

Kühne B, Sprecher E, Enhancement of the production of fungal volatiles employing different adsorbents, Flavour Fragrance J, 1989, **4**, 77

Kwong SCW, Randers L, Rao G, On-line assessment of metabolic activities based on culture redox potential and dissolved oxygen profiles during aerobic fermentation, Biotechnol Prog, 1992, **8**, 576

Larroche C, Gros JB, Characterization of the behavior of *Penicillium roqueforti* in solid state cultivation on support by material balances, J Fermentation Bioeng, 1992, **74**, 305

Larroche C, Theodore M, Gros JB, Growth and sporulation behaviour of *Penicillium roqueforti* solid substrate fermentation - effect of the hydric parameters of the medium, Appl Microbiol Biotechnol, 1992, **38**, 183

Lauritsen FR, Lloyd D, Direct detection of volatile metabolites produced by microorganisms. Membrane inlet mass spectrometry, In: Mass Spectrometry for the Characterization of Microorganisms ACS symp ser 541, ACS Wash DC 1994, 91

Lebo JA, Zajicek JL, Huckins JN, Petty JD, Peterman PH, Use of semipermeable membrane devices for *situ* monitoring of polycyclic aromatic hydrocarbons in aquatic environments, Chemosphere, 1992, **25**, 697

Lethanh M, Voilley A, Tanluu RP, The influence of the composition of model liquid culture medium on vapor liquid partition coefficient of aroma substances, Sci Aliments, 1993, **13**, 699

Lewis VP, Yang ST, A novel extractive fermentation process for propionic acid production from whey lactose, Biotechnol Progress, 1992b, **8**, 104

Lewis VP, Yang ST, Continuous propionic acid fermentation by immobilized *Propionibacterium acidipropionici* in a novel packed-bed bioreactor, Biotechnol Bioeng, 1992a, **40**, 465

Lloyd D, Ellis JE, Hillman K, Williams AG, Membrane inlet mass spectrometry - probing the rumen ecosystem, J Appl Bacteriol, 1992, **73**, 155

Lloyd D, Morrell S, Carlsen HN, Degn H, James PE, Rowlands CC, Effects of growth with ethanol on fermentation and membrane fluidity of *Saccharomyces cerevisiae*, Yeast, 1993, **9**, 825

Lonsane BK, Sancedo-Castaneda G, Raimbault M, Roussos S, Viniegra-Gonzales G, Ghildyal NP, Ramakrishna M, Krishnaiah MM, Scale-up strategies for solid state fermentation systems, Proc Biochem, 1992, **27**, 259

Maia ABRA, Nelson DL, Application of gravitational sedimentation to efficient cellular recycling in continuous alcoholic fermentation, Biotechnol Bioeng, 1993, **41**, 361

Markl H, Zenneck C, Dubach AC, Ogbonna JC, Cultivation of *Escherichia coli* to high cell densities in a dialysis reactor, Appl Microbiol Biotechnol, 1993, **39**, 48

Matsuno R, Adachi S, Uosaki H, Bioreduction of prochiral ketones with yeast cells cultivated in a vibrating air-solid fluidized bed fermentor, Biotechnol Adv, 1993, **11**, 509

Mertooetomo E, Valsaraj KT, Wetzel DM, Harrison DP, Cascade crossflow air stripping of moderatily volatile compounds using high air-to-water ratios, Water Res, 1993, **27**, 1139

Mignone CF, Rossa CA, A simple method for designing fed-batch cultures with linear gradient feed of nutrients, Proc Biochem, 1993, **28**, 405

Mohn WW, Kennedy KJ, Reductive dehalogenation of chlorophenols by *Desulfomonile tiedjei*, Appl Environm Microbiol, 1992, **58**, 1367

Mollah AH, Stuckey DC, Feasibility of *in situ* gas stripping for continuous acetone-butanol fermentation by *Clostridium acetobutylicum*, Enzyme Microbial Technol, 1993, **15**, 200

Morin A, Raymond Y, Cormier F, Production of fatty acid ethyl esters by *Pseudomonas fragi* under conditions of gas stripping, Proc Biochem, 1994, **29**, 437

Najm IN, Snoeyink VL, Richard Y, Removal of 2,4.6-trichlorophenol and natural organic matter from water supplies using PAC in floc-blanket reactors, Water Res, 1993, **27**, 551

Nguyen VT, Shieh WK, Continuous ethanol fermentation using immobilized yeast in a fluidized bed reactor, J Chem Technol Biotechnol, 1992, **55**, 339

Ohara H, Hiraga T, Katasho I, Inuta T, Yoshida T, HPLC monitoring system for lactic acid fermentation, J Fermentation Bioeng, 1993, **75**, 470

Ohrner N, Martinelle M, Mattson A, Norin T, Hult K, Displacement of the equilibrium in lipase catalysed transesterification in ethyl octanoate by continuous evaporation of ethanol, Biotechnol Let, 1992, **14**, 263

Oishi K, Tominaga M, Kawato A, Abe Y, Imayasu S, Nanba A, Development of on-line sensoring and computer aided control systems for sake brewing, J Biotechnology, 1992, **24**, 53

Olsvik ES, Kristiansen B, Influence of oxygen tension, biomass concentration, and specific growth rate on the rheological properties of a filamentous fermentation broth, Biotechnol Bioeng, 1992, **40**, 1293

Ortega GM, Martinez EO, Gonzalez PC, Betancourt D, Otero MA, Enzyme activities and substrate degradation during white rot fungi growth on sugar-cane straw in a solid state fermentation, World J Microbiol Biotechnol, 1993, **9**, 210

Ortiz-Vazquez E, Granados-Baeza M, Rivera-Munoz G, Effect of culture conditions on lipolytic enzyme production by *Penicillium candidum* in a solid state fermentation, Biotechnol Adv, 1993, **11**, 409

Pandey A, Recent process developments in solid-state fermentation, Process Biochem, 1992, **27**, 109

Penaloza W, Davey CL, Hedger JN, Kell DB, Physiological studies on the solid-state quinoa tempe fermentation, using on-line measurements of fungal biomass production, J Sci Food Agric, 1992, **59**, 227

Picque D, Corrieu G, Performances of aseptic sampling devices for on-line monitoring of fermentation processes, Biotechnol Bioeng, 1992, **40**, 919

Pohland HD, Huhn HJ, Klarmann H, Prause M, Soyez K, High performance bacterial fermentation process with 2 gaseous substrates, Acta Biotechnologia, 1992, **12**, 17

Qureshi N, Maddox IS, Friedl A, Application of continuous substrate feeding to the ABE perstraction, stripping, and pervaporation, Biotechnol Progress, 1992, **8**, 382

Rizvi SSH, Yu ZR, Bhaskar A, Rosenberry L, Phase equilibria and distribution coefficients of delta-lactones in supercritical carbon dioxide, J Food Sci, 1993, **58**, 996

Roy PK, Khan AW, Kumar J, Chopra SDK, Basu SK, Steroid transformation in a laboratory-scale glass air-lift fermenter, World J Microbiol Biotechnol, 1992, **8**, 399

Russell AB, Thomas CR, Lilly MD, The influence of vessel height and top-section size on the hydrodynamic characteristics of airlift fermentors, Biotechnol Bioeng, 1994, **43**, 69

Russell JB, Another explanation for the toxicity of fermentation acids at low pH - anion accumulation versus uncoupling, J Appl Bacteriol, 1992, **73**, 363

Sakurai A, Imai H, Effect of operational conditions on the rate of citric acid production by rotating disk contactor using *Aspergillus niger*, J Fermentation Bioeng, 1992, **73**, 251

Samuta T, Ohta T, Ohba T, On-line monitoring of alcohol concentration in sake mash by use of lipqid-membrane-liquid-type membrane distillation and a semiconductor-type alcohol sensor, J Sci Fermentation Technol, 1992, **70**, 191

Sargantanis J, Karim MN, Murphy VG, Ryoo D, Tengerdy RP, Effect of operating conditions on solid substrate fermentation, Biotechnol Bioeng, 1993, **42**, 149

Schlebusch JP PhD Thesis TU Munich 1992

Scholler C, Chaudhuri JB, Pyle DL, Emulsion liquid membrane extraction of lactic acid from aqueous solutions and fermentation broth, Biotechnol Bioeng, 1993, **42**, 50

Schügerl K, Comparison of different bioreactor performances, Bioprocess Eng, 1993, **9**, 215

Schügerl K, Lübbert A, Scheper T, Possibilities to improve the process in food- and biotechnology, Food Biotechnol, 1990, **4**, 241

Shacharnishri Y, Freeman A, Continuous production of acetaldehyde by immobilized yeast with *situ* product trapping - comparison of alcohol dehydrogenase and alcohol oxidase routes, Appl Biochem Biotechnol, 1993, **39**, 387

Smith MR, Vanderschaff A, Deree EM, Debont JAM, Hugenholtz J, The physiology of *Lactococcus lactis* subsp *lactis* biovar *diacetylactis* immobilized in hollow-fibre bioreactors - glucose, lactose and citrate metabolism at high cell densities, Appl Microbiol Biotechnol, 1993, **39**, 94

Stevens S, Hofmeyr JHS, Effects of ethanol, octanoic and decanoic acids on fermentation and the passive influx of protons through the plasma membrane of *Saccharomyces cerevisiae*, Appl Microbiol Biotechnol, 1993, **38**, 656

Strel B, Grba S, Maric V, Enhancement of biomass and fermentation activity of surplus brewers' yeast in a fed-batch process, Appl Microbiol Biotechnol, 1993, **39**, 53

Thomas CR, Image analysis of filamentous fermentation broths, J Chem Technol Biotechnol, 1993, **56**, 204

Tin CSF, Mawson AJ, Ethanol production from whey in a membrane recycle bioreactor, Proc Biochem, 1993, **28**, 217

Turner C, Thornhill NF, Fish NM, A novel method for the on-line analysis of fermentation broth using a sampling device, microcentrifuge and HPLC, Biotechnol Technol, 1993, **7**, 19

Vaccari G, Dosi E, Campi AL, Mantovani G, Proposal for the on-line utilization of the NIR technique to control fermentations, Zuckerindustrie, 1993, **118**, 266

Vaccari G, Gonzalezvara A, Compi AL, Dosi E, Brigidi P, Matteuzzi D, Fermentative production of L-lactic acid by *Lactobacillus casei* DSM-20011 and product recovery using ion exchange resins, Appl Microbiol Biotechnol, 1993, **40**, 23

Vansonsbeek HM, Beeftink HH, Tramper J, 2-liquid-phase bioreactors - review, Enzyme Microbial Technol, 1993, **15**, 722

Villegas E, Aubague S, Alcantara L, Auria R, Revah S, Solid state fermentation - Acid protease production in controlled CO_2 and O_2 environments, Biotechnol Adv, 1993, **11**, 387

Voilley A, Charbit G, Gobert F, Recovery and separation of 1-octen-3-ol from aqueous solutions by pervaporation through silicone membrane, J Food Sci 1990, **55**, 1399

Warren RK, Hill GA, MacDonald DG, Simulation and optimization of a cell recycle system, Appl Biochem Biotechnol, 1992, **34**, 585

Weidenhamer JD, Macias FA, Fischer NH, Williamson GB, Just how insoluble are monoterpenes ?, J Chem Ecol, 1993, **19**, 1799

Weier AJ, Glatz BA, Glatz CE, Recovery of propionic and acetic acids from fermentation broth by electrodialysis, Biotechnol Progress, 1992, **8**, 479

Weusterbotz D, Aivasidis A, Wandrey C, Continuous ethanol production by *Zymomonas mobilis* in a fluidized bed reactor. 2. Process development for the fermentation of hydrolysed B-starch without sterilization, Appl Microbiol Biotechnol, 1993, **39**, 685

Wright JD, Fungal degradation of benzoic acid and related compounds, World J Microbiol Biotechnol, 1993, **9**, 9

Yamamoto K, Ishizaki A, Stanbury PF, Reduction in the length of the lag phase of L-lactate fermentation, J Fermentation Bioeng, 1993, **76**, 151

Zhang S, Matsuoka H, Toda K, Production and recovery of propionic and acetic acids in electrodialysis culture of *Propionibacterium shermanii*, J Fermentation Bioeng, 1993, **75**, 276

Chapter 11

Abraham B, Krings U, Berger RG, Biotechnologische Produktion von Aromastoffen durch Basidiomyceten, GIT Fachz Lab, 1994, **38**, 370

Andersen MR, Omiecinski CJ, Direct extraction of bacterial plasmids from food for polymerase chain reaction amplification, Appl Environm Microbiol, 1992, **58**, 4080

Armstrong DW, Brown LA, Porter S, Rutten R, Biotechnological derivation of aromatic flavour compounds and precursors, In: Progress in Flavour Precursor Studies, Schreier P, Winterhalter P (eds), Allured Carol Stream, IL 1993, 425

Armstrong DW, Gillies B, Yamazaki H, Natural flavors produced by biotechnological processing, In: Flavor Chemistry, Teranishi R, Buttery RG, Shahidi F (eds), ACS symp ser 388, ACS Wash DC 1989, 105

Bardi L, Delloro V, Delfine C, Marzona M, A rapid spectrophotometric method to determine esterase activity of yeast cells in an aqueous medium, J Inst Brewing, 1993, **99**, 385

BFA Laboratories, WO 94/08028, 1994

Bille, PG, Espie WE, Mullan WMA, Evaluation of media for the isolation of leuconostocs from fermented dairy products, Milchwissenschaft - Milk Sci Intern, 1992, **47**, 637

Bowen WR, Sabuni HAM, Ventham TJ, Studies of the cell-wall properties of *Saccharomyces cerevisiae* during fermentation, Biotechnol Bioeng, 1992, **40**, 1309

Braus-Stromeyer SA, Cook AM, Leisinger T, Biotransformation of chloromethane to methanethiol, Environm Sci Technol, 1993, **27**, 1577

Burla G, Garzillo AM, Luna M, Cardelli LE, Schiesser A, Effects of different growth conditions on enzyme production by *Pleurotus ostreatus* in submerged culture, Bioresource Technol, 1992, **42**, 89

Chien MM, Rosazza JP, Microbial transformations of natural antitumor agents: Use of solubilizing agents to improve yields of hydroxylated ellipticines, Appl Environm Microbiol, 1980, **40**, 741

Choi YB, Lee JY, Lim HK, Kim HS, A novel method to supply the inhibitory aromatic compounds in aerobic biodegradation process, Biotechnol Techn, 1992, **6**, 489

Christen P, Raimbault M, Optimization of culture medium for aroma production by *Ceratocystis fimbriata*, Biotechnol Lett, 1991, **13**, 521

Delgado G, Guillen F, Martinez MJ, Gonzalez AE, Martinez AT, Light stimulation of aryl-alcohol oxidase activity in *Pleurotus eryngii*, Mycol Res, 1992, **96**, 984

Dueck G, New optimization problems. The great deluge algorithm and the record-to-record-travel, J Comput Physics, 1993, **104**, 86

Durand A, Renaud R, Almanza S, Maratray J, Diez M, Desgranges C, Solid state fermentation reactors - from lab scale to pilot plant, Biotechnol Adv, 1993, 11, 591

Goetschel R, Barenholz Y, Bar R, Microbial conversions in a liposomal medium. 2. Cholesterol oxidation by *Rhodococcus erythropolis*, Enzyme Microbial Technol, 1992, **14**, 390

Grogan G, Roberts S, Wan P, Willetts A, Camphor-grown *Pseudomonas putida*, a multifunctional biocatalyst for undertaking Baeyer-Villiger monooxygenase-dependent biotransformations, Biotechnol Lett, 1993, **15**, 913

Grogan G, Roberts S, Willetts A, Biotransformations by microbial Baeyer-Villiger monooxygenases stereoselective lactone formation *in vitro* by coupled enzyme systems, Biotechnol Lett, 1992, **14**, 1125

Gross B, Yonnet G, Picque D, Brunerie P, Corrieu G, Aster M, Production of methylanthranilate by the basidiomycete *Pycnoporus cinnabarinus* (Karst.), Appl Microbiol Biotechnol, 1990, **34**, 387

Hanssen HP, Sprecher E, Klingenberg A, Accumulation of volatile flavour compounds in liquid cultures of *Kluyveromyces lactis* strains, Z Naturforsch, 1984, **39c**, 1030

Hariel J, personal commun, 1993, data of 1988/90

Hohler A, Nitz S, Drawert F, Über die Bildung und die sensorischen Eigenschaften flüchtiger Nor-Carotinoide, 1. Photooxidation von β-Carotin in N,N-Dimethylformamid, Chem Mikrobiol Technol Lebensm, 1988, **11**, 115

Huang ZX, Dostal L, Rosazza JPN, Microbial transformations of ferulic acid by *Saccharomyces cerevisiae* and *Pseudomonas fluorescens*, Appl Environm Microbiol, 1993, **59**, 2244

Ivanova N, Yotova L, Biotransformation of furfural by yeast cells covalently bound to cellulose granules, Acta Biotechnolgia, 1993, **13**, 79

Jitsufuchi T, Ishikawa H, Tanaka H, Matsushima K, A simple method of fuzzy modeling for a microorganism reaction, J Fermentation Bioeng, 1992, **74**, 312

Kawakami K, Abe T, Yoshida T, Silicone-immobilized biocatalysts effective for bioconversions in nonaqueous media, Enzyme Microbial Technol, 1992, **14**, 371

Kennedy MJ, Prapulla SG, Thakur MS, Designing fermentation media - a comparison of neural networks to factorial design, Biotech Techn, 1992, **6**, 293

Kodera Y, Furukawa M, Yokoi M, Kuno H, Matsushita H, Inada Y, Lactone synthesis from 16-hydroxyhexadecanoic acid ethyl ester in organic solvents catalyzed with polyethylene glycol-modified lipase, J Biotechnol, 1993, **31**, 219

Kometani T, Yoshii H, Kitatsuji E, Nishimura H, Matsuno R, Large-scale preparation of (S)-ethyl 3-hydroxybutanoate with a high enantiomeric excess through baker's yeast-mediated bioreduction, J Fermentation Bioeng, 1993, **76**, 33

Luus R, Optimization of fed-batch fermentors by iterative dynamic programming, Biotechnol Bioeng, 1993, **41**, 599

Majcherczyk A, Zeddel A, Kelschebach M, Loske D, Huettermann A, Degradation of polycyclic aromatic hydrocarbons by white-rot fungi, BioEngineering, 1993, **9**, 27

Moshy RJ, Impact of biotechnology on food product development, Food Technol, 1985, **39**, 113

Namdev PK, Thompson BG, Ward DB, Gray MR, Effects of glucose fluctuations on synchrony in fed-batch fermentation of *Saccharomyces cerevisiae*, Biotechnol Prog, 1992, 8, 501

Nikolova P, Ward OP, Whole cell yeast biotransformations in 2-phase systems - effect of solvent on product formation and cell structure, J Ind Microbiology, 1992, **10**, 169

Nyttle VG, Chidambaram M, Fuzzy logic control of a fed-batch fermentor, Bioprocess Eng 1993, **9**, 115

Oda S, Ohta H, Microbial transformation on interface between hydrophilic carriers and hydrophobic organic solvents, Biosci Biotech Biochem, 1992, **56**, 2041

Park YS, Shi ZP, Shiba S, Chantal C, Iijima S, Kobayashi T, Application of fuzzy reasoning to control of glucose and ethanol concentrations in baker's yeast culture, Appl Microbiol Biotechnol, 1993, **38**, 649

Pfefferle W, Anke H, Bross M, Steglich W, Inhibition of solubilized chitin synthase by chlorinated aromatic compounds isolated from mushroom cultures, Agric Biol Chem, 1990, **54**, 1381

Pinheiro HM, Cabral JMS, Screening of whole-cell immobilization procedures for the delta-dehydrogenation of steroids in organic medium, Enzyme Microbial Technol, 1992, **14**, 619

Reisman HB, Problems in scale-up of biotechnology production processes, Crit Revs Biotechnol, 1993, **13**, 195

Rütgerswerke DBP-Appl. A 23 L, I/23. OS 4036763, 1992

Sangwan RS, Singhsangwan N, Luthra R, Metabolism of acyclic monoterpenes - Partial purification and properties of geraniol dehydrogenase from lemongrass (*Cymbopogon flexuosus* stapf) leaves, J Plant Physiology, 1993, **142**, 129

Schindler J, Terpenoids by microbial fermentation, Ind Eng Chem Prod Res Dev, 1982, **21**, 537

Schinschel C, Simon H, Preparation of pyruvate from (R)-lactate with proteus species, J Biotechnology, 1993, **31**, 191

Simutis R, Havlik I, Lübbert A, Fuzzy-aided neural network for real-time state estimation and process prediction in the alcohol formation step of production-scale beer brewing, J Biotechnol, 1993, **27**, 203

Taipa MA, Cabral JMS, Santos H, Comparison of glucose fermentation by suspended and gel-entrapped yeast cells - an in vivo nuclear magnetic resonance study, Biotechnol Bioeng, 1993, **41**, 647

Tannock GW, McConnell MA, Fuller R, A note on the use of a plasmid as a RNA probe in the detection of a *Lactobacillus fermentum* strain in porcine stomach contents, J Appl Bacteriol, 1992, **73**, 60

Targonski Z, Biotransformation of lignin-related aromatic compounds by *Pichia stipitis* Pignal, Zentralbl Mikrobiol, 1992, **147**, 244

Teranishi R, New trends and developments in flavor chemistry, In: Teranishi R, Buttery RG, Shahidi F (eds) Flavor Chemistry, ACS symp ser 388, ACS Wash DC 1989, 20

Thibault J, Najim K, Optimization and control of a continuous stirred tank fermenter using learning system, Bioprocess Eng, 1993, **9**, 107

Ushio K, Hada J, Tanaka Y, Ebara K, Allyl bromide, a powerful inhibitor against R-enzyme activities in bakers' yeast reduction of ethyl 3-oxoalkanoates, Enzyme Microbial Technol, 1993, **15**, 222

van der Schaft PH, ter Burg N, van den Bosch S, Cohen AM, Fed-batch production of 2-heptanone by *Fusarium poae*, Appl Microbiol Biotechnol, 1992, **36**, 709

Vermue M, Sikkema J, Verheul A, Bakker R, Tramper J, Toxicity of homologous series of organic solvents for the gram-positive bacteria *Arthrobacter* and *Nocardia* sp and the gram-negative bacteria *Acinetocacter* and *Pseudomonas* sp, Biotechnol Bioeng, 1993, **42**, 747

Votruba J, Sobotka M, Physiological similarity and bioreactor scale-up, Folia Microbiol, 1992, **37**, 331

Vulfson EN, Enzymatic synthesis of food ingredients in low-water media, Trends Food Sci Technol, 1993, **4**, 209

Wang HH, Assessment of solid state fermentation by a bioelectronic artificial nose, Biotechnol Adv, 1993, **11**, 701

Wolfram JH, Rogers RD, Higdem DM, Microbial processing of volatile organics in industrial waste streams, J Environm Sci and Health A, 1992 **A27**, 1115

Zimmermann HJ, Fuzzy set theory- and its application, 2nd ed, Kluwer, Dordrecht, 1991

Zuniga M, Pardo I, Ferrer S, An improved medium for distinguishing between homofermentative and heterofermentative lactic acid bacteria, Intern J Food Microbiol, 1993, **18**, 37

Chapter 12

Aymerich MT, Hugas M, Garriga M, Vogel RF, Monfort JM, Electrotransformation of meat lactobacilli - effect of several parameters on their efficiency of transformation, J Appl Bacteriol, 1993, **75**, 320

Buck L, Identification and analysis of a multigene family encoding odorant receptors: implications for mechanisms underlying olfactory information processing, Chem Senses, 1993, **18**, 203

Bussolati L, Ramoni R, Grolli S, Donofrio G, Bignetti E, Preparation of an affinity resin for odorants by coupling odorant binding protein from bovine nasal mucosa to sepharose 4B, J Biotechnology, 1993, **30**, 225

Columbia University, Odorant receptors, cDNA encoding the receptors, and uses thereof, WO 9217585, 1992

Croteau R, The biosynthesis of thujane monoterpenes, In: Recent Developments in Flavor and Fragrance Chemistry, Hopp R, Mori K (eds), 1993, VCH, 261

Ganeva VJ, Tsoneva IC, Effect of N-alcohols on the electrotransformation and permeability of *Saccharomyces cerevisiae*, Appl Microbiol Biotechnol, 1993, **38**, 795

Graziadei PPC, Olfactory organ culture in vivo and in vitro, Cytotechnology, 1993, **11**, 3

Hartmeier W, Coimmobilization of enzymes and whole cells, Food Biotechnol, 1990, **4**, 399

Hidaka T, Omura M, Transformation of *Citrus* protoplasts by electroporation, J Jap Soc Horticult Sci, 1993, **62**, 371

John Hopkins University, Primary culture of olfactory neurons, US 5217893 A 930608, 1993

Lehn C, Freeman A, Schuhmann W, Schmidt HL, Stabilization of NAD^+-dependent dehydrogenases and diaphorase by bilayer encagement, J Chem Technol Biotechnol, 1992, **54**, 215

Otte S, Etherische Öle-Wiederentdeckte Heilmittel, Dragoco Bericht, 1994, **3**, 91

Querol A, Barrio E, Huerta T, Ramon D, Molecular monitoring of wine fermentations conducted by active dry yeast strains, Appl Environm Microbiol, 1992, **58**, 2948

Raming K, Krieger J, Strotmann J, Boekhoff I, Kubick S, Baumstark C, Breer H, Cloning and expression of odorant receptors, Nature, 1993, **361**, 353

Roig MG, Kennedy JF, Perspectives for chemical modifications of enzymes, Crit Revs Biotechnol, 1992, **12**, 391

Ronnett GV, Cho H, Hester LD, Wood SF, Snyder SH, Odorants differentially enhance phosphoinositide turnover and adenylyl cyclase in olfactory receptor neuronal cultures, J Neurosci, 1993, **13**, 1751

Russell C, Jarvis A, Yu PL, Mawson J, Optimization of an electroporation procedure for *Kluyveromyces lactis* transformation, Biotechnol Techn, 1993, **7**, 417

Schleifer KH, DNA probes in food microbiology, Food Biotechnology, 1990, **4**, 585

Schreier P, Winterhalter P (eds) 1992, Progress in Flavour Precursor Studies, Allured Carol Stream IL

Shepherd GM, Current issues in the molecular biology of olfaction, Chem Senses, 1993, **18**, 191

Stredansky M, Svarc R, Sturdik E, Dercova K, Repeated batch alpha-amylase production in aqueous 2-phase system with *Bacillus* strains, J Biotechnology, 1993, **27**, 181

Teranishi R, Takeoka GR, Güntert M (eds) 1992, Flavor Precursors, ACS symp ser 490, ACS Wash DC

Wiseman A, Designer enzyme and cell applications in industry and in environmental monitoring, J Chem Technol Biotechnol, 1993, **56**, 3

Yang FX, Russell AJ, Optimization of baker's yeast alcohol dehydrogenase activity in an organic solvent, Biotechnol Prog, 1993, **9**, 234

Index

Auf dem neuesten Stand der Technik

E. Lück, M. Jager

Chemische Lebensmittelkonservierung

Stoffe, Wirkungen, Methoden
3., überarbeitete Auflage 1995. Etwa 250 Seiten.
8 Abbildungen, 36 Tabellen.
Gebunden ca. DM 168,–   ISBN 3-540-57607-X
Erscheint August 1995 (voraussichtlich)

Die Autoren dieses Standardwerks haben für die 3. Auflage alle
Kapitel grundlegend überarbeitet und dem neuesten Erkenntnis-
stand angepaßt; zahlreiche neue Konservierungsstoffe wurden
berücksichtigt. Das erfolgreiche Grundkonzept wurde beibehal-
ten: Ein allgemeiner Teil führt in die wesentlichen Grundlagen
der chemischen Lebensmittelkonservierung ein; im speziellen
Teil werden die einzelnen Konservierungsstoffe, ihre Eigenschaften und Anwendungsgebiete behandelt. Das Ziel der
Autoren ist dabei stets, Grundlagen für die praktische Anwen-
dung zu vermitteln und Zusammenhänge erkennbar werden zu
lassen. Auf die historische Entwicklung und international
bedeutsame Tendenzen wird der Leser ebenfalls hingewiesen.

Springer

Preisänderungen vorbehalten.
In Ländern der EU gilt die landesübliche MWSt.

Springer-Verlag, P. O. Box 31 13 40, D-10643 Berlin, Germany. Fax: (0)30/82 07-301/448, e-mail: orders@springer.de

Printing: Mercedesdruck, Berlin
Binding: Buchbinderei Lüderitz & Bauer, Berlin